COMPUTATIONAL INTELLIGENCE IN MANUFACTURING

Woodhead Publishing Reviews:
Mechanical Engineering Series

COMPUTATIONAL INTELLIGENCE IN MANUFACTURING

Edited by

KAUSHIK KUMAR

GANESH KAKANDIKAR

J. PAULO DAVIM

WOODHEAD
PUBLISHING

ELSEVIER   An imprint of Elsevier

Woodhead Publishing is an imprint of Elsevier
50 Hampshire Street, 5th Floor, Cambridge, MA 02139, United States
The Boulevard, Langford Lane, Kidlington, OX5 1GB, United Kingdom

Notices

Knowledge and best practice in this field are constantly changing. As new research and experience broaden our understanding, changes in research methods, professional practices, or medical treatment may become necessary.

Practitioners and researchers must always rely on their own experience and knowledge in evaluating and using any information, methods, compounds, or experiments described herein. In using such information or methods they should be mindful of their own safety and the safety of others, including parties for whom they have a professional responsibility.

To the fullest extent of the law, neither the Publisher nor the authors, contributors, or editors, assume any liability for any injury and/or damage to persons or property as a matter of products liability, negligence or otherwise, or from any use or operation of any methods, products, instructions, or ideas contained in the material herein.

ISBN: 978-0-323-91854-1 (print)

ISBN: 978-0-323-91855-8 (online)

For information on all Woodhead publications
visit our website at https://www.elsevier.com/books-and-journals

Publisher: Matthew Deans
Acquisitions Editor: Brian Guerin
Editorial Project Manager: Rafael Guilherme Trombaco
Production Project Manager: Surya Narayanan Jayachandran
Cover Designer: Mark Rogers

Typeset by STRAIVE, India

Working together
to grow libraries in
developing countries

www.elsevier.com • www.bookaid.org

Contents

Contributors

Sachin Ambade
Department of Mechanical Engineering, Yeshwantrao Chavan College of Engineering, Nagpur, India

Maithili Anil Chougule
School of Mechanical Engineering, Dr Vishwanath Karad MIT World Peace University, CAD/CAM/CAE, Pune, India

Kumar Anuj
Apeejay School of Management, Delhi, India

Buttan Apoorva
Amity School of Communication, Amity University, Noida, India

Ratnakar Ghorpade
School of Mechanical Engineering, Dr. Vishwanath Karad MIT World Peace University, Pune, India

Vikas Gohil
Mechanical Engineering Department, Bajaj Institute of Technology, Pipri, Wardha, India

S. Gowri
Department of Manufacturing Engineering, College of Engineering Guindy, Anna University, Chennai, India

P. Hariharan
Department of Manufacturing Engineering, College of Engineering Guindy, Anna University, Chennai, India

Pradeep Jadhav
Bharati Vidyapeeth (Deemed to be University) College of Engineering, Mechanical Engineering Department, Pune, India

Balram Jaiswal
Materials and Morphology Laboratory, Department of Mechanical Engineering, Madan Mohan Malaviya University of Technology, Gorakhpur, Uttar Pradesh, India

Sabharwal Jyotsana
Apeejay School of Management, Delhi, India

Ganesh M. Kakandikar
School of Mechanical Engineering, Dr. Vishwanath Karad MIT World Peace University, Pune, India

M. Kathiresan
Department of Mechanical Engineering, E.G.S. Pillay Engineering College, Nagapattinam, India

Navneet Khanna
Advanced Manufacturing Laboratory, Institute of Infrastructure Technology Research and Management (IITRAM), Ahmedabad, India

Nitin Kotkunde
Department of Mechanical Engineering, BITS Pilani, Hyderabad, India

Omkar K. Kulkarni
School of Mechanical Engineering, Dr. Vishwanath Karad MIT World Peace University, Pune, India

Jogendra Kumar
Materials and Morphology Laboratory, Department of Mechanical Engineering, Madan Mohan Malaviya University of Technology, Gorakhpur, Uttar Pradesh, India

Kaushlendra Kumar
Materials and Morphology Laboratory, Department of Mechanical Engineering, Madan Mohan Malaviya University of Technology, Gorakhpur, Uttar Pradesh, India

Kuldeep Kumar
Materials and Morphology Laboratory, Department of Mechanical Engineering, Madan Mohan Malaviya University of Technology, Gorakhpur, Uttar Pradesh, India

Anil S. Mashalkar
School of Mechanical, Dr. Vishwanath Karad MIT World Peace University, Pune, India

Diwesh Babruwan Meshram
Department of Plastics Engineering, Central Institute of Petrochemicals Engineering and Technology, Korba, India

Arora Monika
Apeejay School of Management, Delhi, India

Ayush Morchhale
Department of Mechanical Engineering, BITS Pilani, Hyderabad, India

Vilas M. Nandedkar
Production Engineering Department, Shri Guru Gobind Singhji Institute of Engineering and Technology, Nanded, India

Ganadhar Rajaram Navnage
Department of Metallurgical Engineering and Materials Science, VNIT, Nagpur, India

Anita Nene
Dr. Vishwanath Karad MIT World Peace University, Pune, India

Swanand Pachpore
Bharati Vidyapeeth (Deemed to be University) College of Engineering, Mechanical Engineering Department; School of Mechanical Engineering, Dr. Vishwanath Karad MIT World Peace University, Pune, India

P. Pandiarajan
Department of Mechanical Engineering, Theni Kammavar Sangam College of Technology, Theni, India

Sandeep Pandre
Department of Mechanical Engineering, BITS Pilani, Hyderabad, India

Yogesh M. Puri
Department of Mechanical Engineering, VNIT, Nagpur, India

Pujari Purvi
Bharati Vidyapeeth's Institute of Management Studies and Research, Mumbai, India

Ambuj Saxena
G L Bajaj Institute of Technology and Management, Greater Noida, India

Kesarwani Shivi
Materials and Morphology Laboratory, Department of Mechanical Engineering, Madan Mohan Malaviya University of Technology, Gorakhpur, Uttar Pradesh, India

Devendra Kumar Singh
Materials and Morphology Laboratory, Department of Mechanical Engineering, Madan Mohan Malaviya University of Technology, Gorakhpur, Uttar Pradesh, India

Swadesh Kumar Singh
Department of Mechanical Engineering, GRIET, Hyderabad, India

P. Sivaprakasam
Department of Mechanical Engineering, College of Electrical and Mechanical Engineering, Center of Excellence - Nano Technology, Addis Ababa Science and Technology University, Addis Ababa, Ethiopia

R. Theerkka Tharisanan
Department of Mechanical Engineering, Theni Kammavar Sangam College of Technology, Theni, India

J. Udaya Prakash
Department of Mechanical Engineering, Vel Tech Rangarajan Dr. Sagunthala R & D Institute of Science and Technology, Chennai, India

Rajesh Kumar Verma
Materials and Morphology Laboratory, Department of Mechanical Engineering, Madan Mohan Malaviya University of Technology, Gorakhpur, Uttar Pradesh, India

Rahul Vishwakarma
Materials and Morphology Laboratory, Department of Mechanical Engineering, Madan Mohan Malaviya University of Technology, Gorakhpur, Uttar Pradesh, India

Preface

We are pleased to present the book *Computational Intelligence in Manufacturing: Industry 4.0* as a part of the Woodhead Publishing Reviews: Mechanical Engineering Series. The title of this book was chosen taking into account the current and growing importance of *computational intelligence* and recognizing its application to an important domain of the industrial and manufacturing world, "Industry 4.0." Thus, this book is an amalgamation of three different domains, namely, *computer intelligence, manufacturing*, and *Industry 4.0*, which makes this book unique.

Computational intelligence is the theory, design, application, and development of biologically and linguistically motivated computational paradigms. Traditionally, the three main pillars of computational intelligence have been neural networks, fuzzy systems, and evolutionary computation. However, with time, many nature-inspired computing paradigms have evolved. Thus, computational intelligence is an evolving field, and at present in addition to the three main constituents, it encompasses computing paradigms like ambient intelligence, artificial life, cultural learning, artificial endocrine networks, social reasoning, and artificial hormone networks. Computational intelligence plays a major role in developing successful intelligent systems.

Manufacturing is the most important field of engineering by virtue of which everything is created. Computational intelligence can be of great help in solving many complicated problems in the manufacturing domain for Industry 4.0. Machine tools need to be designed for automation and agility. Manufacturing cells with flexibility and enhanced productivity and better supply chain management are key for success. There is a lot of ongoing research in the manufacturing processes itself. There is a demand to design and develop specialized processes for newer materials such as composites and alloys. Miniaturization has encouraged industry to develop micro- and nano-manufacturing processes.

Industry 4.0 refers to a new phase in the Industrial Revolution that focuses heavily on interconnectivity, automation, applications of machine learning, and real-time data analysis in manufacturing. It is also known as smart manufacturing of digital manufacturing. Manufacturing processes and systems are going through rapid changes to accommodate the need of the hour in globalized competition and customer-driven product development.

The complete manufacturing system in light of Industry 4.0 is changing rapidly. The pace at which technology is changing is also very quick. Additive manufacturing is changing the way things are produced, opening up a new regime. To obtain maximum benefits from plant resources, flexible manufacturing and agile manufacturing are preferred. New products designed for the global scenario for vibrant customers generate enormous amount of data, which need to be handled by techniques like Big Data. Automation and robotics have led to enhancements in productivity and efficiency. New concepts like swarm robotics can be easily implemented on the shop floor for assembly operations. The equipment used for material handling have become smarter (e.g., automated guided vehicles). Condition monitoring and maintenance of high-end equipment is a major challenge. Computational intelligence can help in the early identification of any probability of breakdown.

This book is focused on all facets of manufacturing processes and systems in the context of Industry 4.0 and is targeted to cater to the needs of students, researchers and industry practitioners, engineers, and research scientists/academicians involved in the development and use of computational intelligence and associated manufacturing technologies to achieve the objectives of Industry 4.0. The book has 10 chapters catering to different aspects of the targeted scope.

Chapter 1 discusses multi-verse multi-objective optimization of thinning and wrinkling in automotive connectors. The authors have taken up one of the most versatile forming processes—sheet metal-forming process. Computer-aided numerical experimentation has been carried out with experiments designed to understand the parameters that influence thinning and wrinkling.

Chapter 2 presents the use of computer intelligence in machining a curved cooling hole in plastic injection molds. The target was to maximize the material removal rate with minimum tool wear in a nontraditional computer-controlled machine (i.e., electrical discharge machining machine). The chapter not only provides the optimal parameters but also evaluates dimensional accuracy along with the targeted outputs.

Chapter 3 investigates the deformation and forming behaviors of dual-phase steel (DP 590) at elevated temperatures. Again, sheet metal processing has been targeted and different process parameters like strain, strain rate, and temperature have been considered to study the flow stress behavior of the material during forming. The flow behavior of DP 590 steel at the tested temperatures and strain rates is predicted using different constitutive and artificial neural network (ANN) models and a comparative analysis has been done.

In Chapter 4, the slime mold algorithm is used to optimize the thermal efficiency of a Scheffler solar concentrator receiver. The Scheffler concentrator is a fixed-focus elliptical dish system used to harvest solar energy. It can concentrate the sunlight to a fixed point where the receiver is mounted. The focused sunlight is used to heat a suitable fluid contained in the receiver. Generally, a cylindrical-shaped receiver is used with the Scheffler collector system. In this chapter, four conditions of the cylindrical receiver were considered for experimentation to generate steam. The slime mold algorithm has been applied to create the regression equation for thermal efficiency so that predictive analysis can be done to determine the best results before installing such huge systems.

In the present era, polymeric composites are the most sought after materials due to their lower weight-to-volume ratio. Chapter 5 studies the drilling behavior of polymeric hybrid composites. The authors have analyzed the drilling parameters, such as spindle speed, feed, filler content, of carbon nanofiller/fiber-reinforced polymer hybrid composites. A metaheuristic nature–inspired algorithm, the ant lion optimization (ALO) algorithm, has been applied using MATLAB® codes. The results of the work can be applied to control the production process quality and productivity indices dealing with varying constraints and multiple responses.

Chapter 6 provides a detailed machining performance analysis of a titanium alloy using a micro-ED milling process. The properties like exceptional hardness and work hardening made the alloy difficult to cut using traditional machining, but its good electrical conductivity has made it one of the most suitable materials for the micro-ED milling process. The chapter describes the use of the response surface methodology to analyze the effect of input factors on the micro-ED milling of a titanium alloy for optimal responses and also provides a regression relationship for prediction and estimation.

Chapter 7 studies the computational analysis of the white layer properties of aluminum metal matrix composites by electrical discharge machining (EDM; both spark and wire types). The highly accelerated heat flux on machined surfaces in EDM and wire EDM causes thermal damage in the heat-affected zone (HAZ), namely, white layer or recast layer and globule. The white layer has high strength and higher hardness, leading to its failure.

The chapter provides a large amount of data on failures and also a solution to the problem.

Chapter 8 presents a discussion on the scope of Industry 4.0 components in small and medium-sized enterprises (SMEs) in the manufacturing sector. The chapter focuses on recent technologies that are creating an impact in SMEs using computational intelligence or soft intelligence. Computational intelligence plays a pivotal role in developing applications and creating algorithms for forecasting market situations and growth by applying different analytical methods like fuzzy logic, neural networks, evolutionary computation, learning theory, and probability methods to increase the productivity and efficiency in SMEs. The chapter elaborately deals with these and provides insight into the applications to achieve the targets under the umbrella of Industry 4.0.

Chapter 9 elaborates the process parameter optimization for manufacturing a root canal device, a biomedical application. Root canal treatment is a therapeutic intervention that aids in the recovery of a decaying and sore tooth and is used to preserve diseased or partially destroyed teeth. This chapter proposes a mathematical model to understand the nature of sliding friction and optimize it with theoretical bounds using the Gorilla Troops optimization algorithm.

Chapter 10 provides a comprehensive review of agricultural irrigation using artificial intelligence for improvement of crop productivity. This chapter deals with the application of various artificial intelligence techniques in agriculture for the improvement of soil fertility and optimization of water used to help improve productivity, quality, and efficiency. The chapter concludes with the use of artificial intelligence technology to develop systems/tools that can predict or prevent issues that could occur in the future, which would make agricultural systems more efficient, thus saving money and time.

First and foremost, we thank God, whose blessings ensured that this work could be completed to our satisfaction despite the difficult times. You have given us the power to believe in passion and hard work and pursue our dreams. We could not have undertaken this herculean task without the faith we have in you, the Almighty. We are thankful to you for this.

We thank all the chapter contributors, the reviewers, editorial advisory board members, book development editor, and the team at Elsevier for being available to work on this project.

We also thank our colleagues, friends, and students at our home institutes and from different institutes and organizations. They have always supported us throughout our careers and while writing this book. This book was not only inspired by them, it was also directly improved by their active involvement in its development. We look forward to discussing this book with them at future gatherings, as we are sure they will all read this book soon.

Throughout the process of editing this book, many individuals, from different walks of life, have taken time out to help. Last, but definitely not least, we thank all of them and our well-wishers for encouraging us. We would have probably given up without their support.

Kaushik Kumar
Ganesh Kakandikar
J. Paulo Davim

Multiverse multiobjective optimization of thinning and wrinkling in automotive connector

Ganesh M. Kakandikar[a], Vilas M. Nandedkar[b], and Omkar K. Kulkarni[a]
[a]School of Mechanical Engineering, Dr. Vishwanath Karad MIT World Peace University, Pune, India
[b]Production Engineering Department, Shri Guru Gobind Singhji Institute of Engineering and Technology, Nanded, India

1.1 Sheet-metal forming

Sheet-metal forming is a significant manufacturing process for producing a large variety of automotive parts (body panels, fenders, etc.) and aerospace parts (body panels, wing parts, etc.), as well as consumer products (kitchen sinks, cans, boxes, etc.) (Ahmetoglu, Broek, Kinzel, & Altan, 1995). The forming process is broadly classified into forming/drawing/stamping, and deep drawing operations, which include a wide spectrum of operations and flow conditions. Deep drawing is a compression-tension forming process with the greatest range of applications, involving rigid tooling, draw punches, a blank holder, and a female die (Bleck, Deng, Papamantellos, & Gusek, 1998). The blank is generally pulled over the draw punch into the die; the blank holder prevents wrinkling from taking place in the flange (Obermeyer & Majlessi, 1998). There is great interest in the formability of metals due to the continuous demand on industry to produce lighter weight yet stronger and more rigid components.

1.2 Automotive component under study: Connector

The connector under study is manufactured by Vishwadeep Enterprises, Pune, for Tata Engineering and Locomotive Co. Ltd., Pune. This connector is fitted between the fuel neck and filler nose. The connector has a total height of 18 mm with the upper 08-mm portion having a 48-mm diameter, and the lower 08-mm step having a 31-mm diameter. The connecting portion has a 2-mm height with 2.5-mm and 3-mm connecting

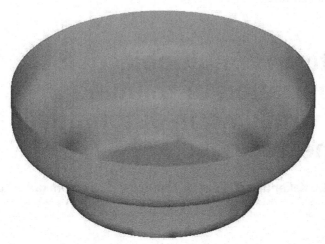

Fig. 1.1 Connector—Geometric model. *No permission required.*

radii, as shown in Fig. 1.1. The details of the connector are given in Table 1.1.

1.3 Taguchi design of experiments

To identify the influence of blank holder force, coefficient of friction, die profile radius, and punch nose radius on thinning and wrinkling, the Taguchi design of experiments was applied. Every process variable was considered with three levels for operation: low, medium, and high. The orthogonal array selected for this combination of 04 parameters and 03 levels is L9. Table 1.2 shows the parameters and their three levels. Table 1.3 presents the detailed L9 orthogonal array.

Table 1.1 Connector—Details.

Manufactured by	Vishwadeep Enterprises, Pune
Component of	Tata Engineering and Locomotive Company Co. Ltd.
Part No.	2779 4710 82 08
Weight	20 g
Material	D-513, SS 4010
Thickness	1.00 mm
Yield Strength	280 MPa
Ultimate Tensile Strength	270–410 MPa
r	1.7 min
n	0.22

No permission required.

Table 1.2 Connector—Levels of process parameters.

	Lower	Middle	Higher
BHF	04 KN	05 KN	06 KN
μ	0.05	0.10	0.15
RD	2.5 mm	3.0 mm	3.5 mm
RP	8.0 mm	9.0 mm	10.0 mm

Table 1.3 Connector—L9 orthogonal array.

Expt. no.	Blank holder pressure	Coefficient of friction	Die cushion radius	Punch nose radius
01	04	0.05	2.5	8.0
02	04	0.10	3.0	9.0
03	04	0.15	3.5	10.0
04	05	0.05	3.0	10.0
05	05	0.10	3.5	8.0
06	05	0.15	2.5	9.0
07	06	0.05	3.5	9.0
08	06	0.10	2.5	10.0
09	06	0.15	3.0	8.0

1.4 Numerical simulation results—Connector

Nine experiments were conducted as per the experiment design in L9. The average blank diameter calculated by the software is 65.93 mm and the average punch force is 54,600 N, as shown in Fig. 1.2, whereas the analytically calculated blank has a diameter of 65 mm. Fig. 1.3 shows the blank

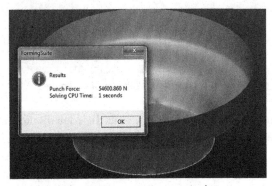

Fig. 1.2 Connector—Punch force. *No permission required.*

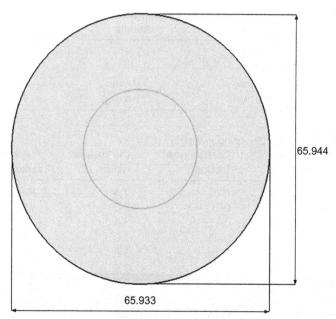

Fig. 1.3 Connector—Blank. *No permission required.*

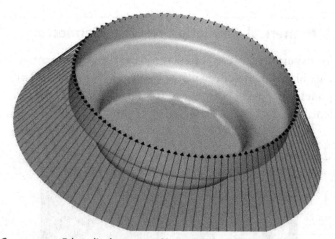

Fig. 1.4 Connector—Edge displacement. *No permission required.*

predicted by the software. Fig. 1.4 represents the edge displacement from the blank to the required shape of the connector. Flow lines represent the flow of material and the path traveled to be converted into the required form with the help of the die and punch. The forming suite is applied for numerical modeling and simulation. It is one of the popular platforms for

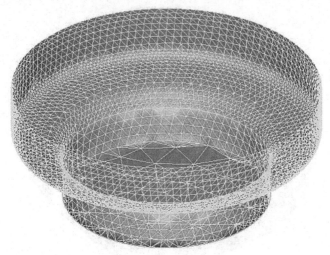

Fig. 1.5 Connector—Meshing. *No permission required.*

metal-forming analysis. Discretization of the component is shown in Fig. 1.5, with mixed elements applied to deal with error. Fig. 1.6 shows the principal curvature directions during forming operations and Fig. 1.7 shows the principal strain directions both in tensile and compressive directions, in red and blue colors, respectively.

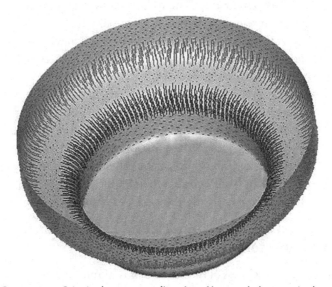

Fig. 1.6 Connector—Principal curvature direction. *No permission required.*

Fig. 1.7 Connector—Principal strain direction. *No permission required.*

1.5 Connector—Forming zone

To gain insight into forming in the connector manufacturing, a forming zone observed in all nine experiments was examined. The bottom flat area has a tight panel in red. Plain strain is observed near the bottom area in certain patches, shown in orange. The walls have a semitight panel, as shown in yellow. The green region shows loose material, which is observed in the upper step. The violet region indicates wrinkling tendency, as the flange has strong wrinkling behavior due to high compressive stresses (Moshksar & Mansorzadeh, 2003), shown in Fig. 1.8.

The bottom part in the second stage shows a tight panel. The wall region of the second stage has a semitight panel with a tight panel at a few nodes. For the upper stage of the connector, the flat bottom indicates loose material with a rim of semitight panel at the junction of the flat bottom and wall. The loose material region extends to a certain height in the upper stage wall region. Above this, the wall shows a wrinkling tendency. Plain strain is observed in the wall region of the lower stage, as shown in Fig. 1.9.

1.6 Connector—Thickness distribution

The original thickness of the connector is 01 mm. It was observed during the L9 array experimentation that there was thinning as well as wrinkling

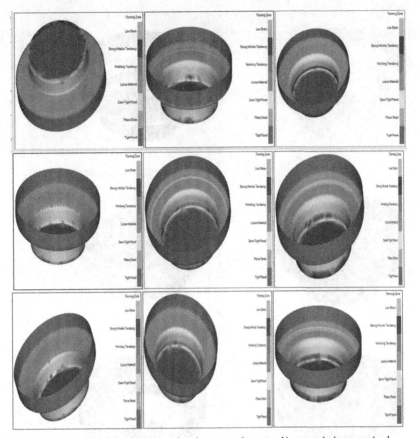

Fig. 1.8 Connector—Forming zone in nine experiments. *No permission required.*

behavior. The thickness ranges for nine experiments are presented in Table 1.4. The thickness distributions in nine experiments are shown in Fig. 1.10.

The thickness variation ranges from 0.81 mm to 1.13 mm in the different regions of the tail cap. The thickness at the bottom varies between 0.87 mm to 0.89 mm. The wall region of the lower stage has a thickness variation between 0.83 mm to 0.87 mm. In the flat bottom of the upper stage, the thickness increases between 0.96 mm to 0.98 mm, whereas the lower portion of the wall of the upper stage has a thickness between 0.98 mm to 1.00 mm. The highest thickness recorded is at the top of the wall and ranges between 1.11 mm to 1.13 mm. The thickness distribution is shown in Fig. 1.11.

Fig. 1.9 Connector—Forming zone at different cross sections. *No permission required.*

Table 1.4 Connector—Thickness distribution.

Expt. I	0.81–1.13 mm	Expt. II	0.78–1.13 mm	Expt. III	0.59–0.89 mm
Expt. IV	0.65–0.89 mm	Expt. V	0.62–0.89 mm	Expt. VI	0.56–0.89 mm
Expt. VII	0.66–0.90 mm	Expt. VIII	0.61–0.89 mm	Expt. IX	0.57–0.89 mm

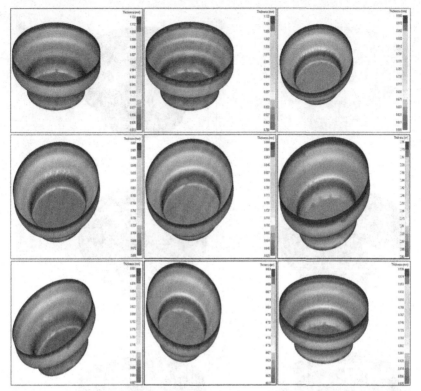

Fig. 1.10 Connector—Thickness distribution in nine experiments. *No permission required.*

1.7 Connector—Safety zone

The safety zone for all nine experiments conducted is plotted in Fig. 1.12. In all experiments the bottom region and most of the wall side region of the lower and upper stage indicates the safe zone (Ghoo & Keum, 2000; Zimniak, 2000). There are also patches of marginal elements. In Experiments 6 and 9, a failure region is observed at the junction of the lower and upper stage. There is no low strain region; a strong wrinkling tendency is also observed, as in Fig. 1.12. The upper wall region shows a wrinkling tendency (Kim & Son, 2000; Lei, Hwang, & Kang, 2001).

1.8 Connector—Safety margin

The plots for safety margin for all nine experiments are presented in Fig. 1.13. It is observed that safety margin is highest at the bottom of the connector and in some experiments at the upper region of the wall.

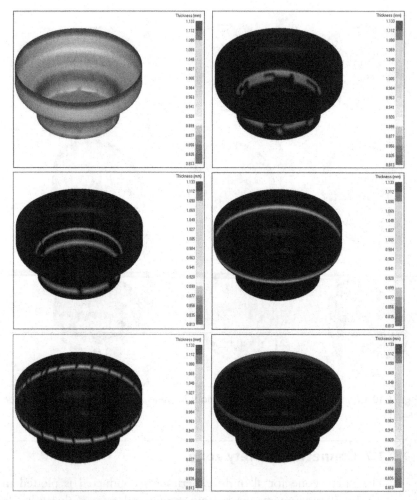

Fig. 1.11 Connector—Thickness distribution. *No permission required.*

It decreases slowly toward the wall region and is found to be lowest at the junction of the lower and upper stage (Chan, Chew, Lee, & Cheok, 2004; Kawka, Olejnik, Rosochowski, Sunaga, & Makinouchi, 2001; Naceur, Guo, Batoz, & Knopf-Lenoir, 2001). The highest safety margin is 0.87 and the lowest is −0.19. In a few experiments, the safety margin is observed as negative (Xia, Shima, Kotera, & Yasuhuku, 2005). The range of safety margin for all experiments is shown in Table 1.5. The connector safety margins for the nine experiments are shown in Fig. 1.13.

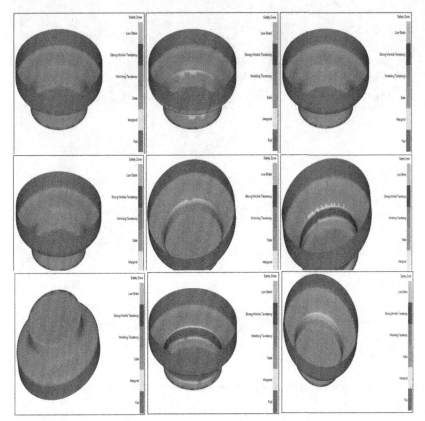

Fig. 1.12 Connector—Safety zone in nine experiments. *No permission required.*

1.9 Connector—Forming limit diagram

The forming limit diagrams (Obermeyer & Majlessi, 1998; Ozturk & Lee, 2004) have been plotted for all nine experiments for the connector. Experiments 3, 6, 8, and 9 show failure points. There are also points that show a wrinkling tendency. The maximum points are in the safe region as well as in the marginal area. The failures occur at 34.33% and 37.12% of major strain. The combination of maximum major and minor strain is presented in Table 1.6, and the forming limit diagrams for all nine experiments are shown in Fig. 1.14.

1.10 Analysis of variance for thinning—Connector

The decrease in the thickness at various cross sections during all nine experiments is measured and presented in Table 1.7. Thickness is measured

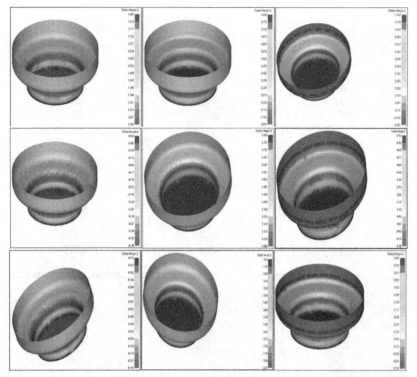

Fig. 1.13 Connector—Safety margin in nine experiments. *No permission required.*

at various cross sections and the average thickness is calculated (Ayed & Batoz, 2005). For analysis of variance, the quality characteristic selected was the difference between original thickness and decreased thickness. The S/N ratios were calculated for a quality characteristic of "smaller is better."

The minimum thickness observed is 0.637 mm. The average decrease in thickness for Experiment 1 is minimal, at 0.108 mm. The analysis of variance results are presented in Tables 1.8 and 1.9. The mean S/N ratios were calculated for all parameters: blank holder force, coefficient of friction, die profile radius, and punch nose radius at all three levels of low, medium, and

Table 1.5 Connector—Safety margin range in nine experiments.

Expt. I	0.16–0.86	Expt. II	0.07–0.82	Expt. III	−0.08–0.76
Expt. IV	0.14–0.86	Expt. V	0.05–0.80	Expt. VI	−0.19–0.76
Expt. VII	0.18–0.87	Expt. VIII	−0.04–0.80	Expt. IX	−0.15–0.76

No permission required.

Table 1.6 Major and minor strain for nine experiments.

Expt. I	−37.2 and 63.1	Expt. II	−38.8 and 67.6	Expt. III	−41.7 and 76.2
Expt. IV	−36.5 and 61.1	Expt. V	−38.6 and 66.9	Expt. VI	−44.5 and 85.8
Expt. VII	−35.4 and 58.2	Expt. VIII	−39.4 and 69.4	Expt. IX	−43.0 and 80.6

high. The range is defined as the difference between the maximum and minimum value of S/N ratio for the particular parameter.

Table 1.9 shows the rearrangement of S/N ratios for all variables for all levels. The rank indicates the influence of the input parameter on the quality characteristic. The result for this orthogonal array indicates that the blank holder force has a major influence on the decrease in thickness. Friction has second place, punch nose radius has third place, and the die profile radius has the least influence on thickness reduction.

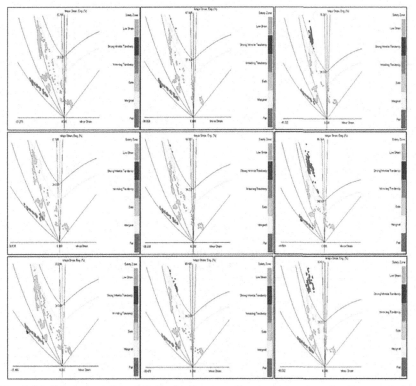

Fig. 1.14 Connector—Forming limit diagrams in nine experiments. *No permission required.*

Table 1.7 Connector—S/N ratios in nine experiments for thinning.

Expt. no.	Decreased thickness (mm)					Average thickness (mm)	Thickness difference (mm)	S/N ratio
1	0.952	0.888	0.824	0.866	0.845	0.8913	0.108	19.27
2	0.932	0.909	0.885	0.862	0.838	0.8735	0.126	17.95
3	0.862	0.842	0.822	0.801	0.781	0.7546	0.245	12.20
4	0.873	0.857	0.841	0.825	0.809	0.782	0.218	13.23
5	0.709	0.691	0.673	0.655	0.637	0.761	0.239	12.43
6	0.862	0.840	0.817	0.795	0.773	0.7441	0.255	11.84
7	0.877	0.862	0.846	0.815	0.800	0.7839	0.216	13.30
8	0.865	0.846	0.828	0.790	0.771	0.7511	0.248	12.08
9	0.863	0.842	0.799	0.778	0.756	0.7415	0.258	11.75

Table 1.8 Connector—S/N ratios at three levels for thinning.

Parameter	Level	Experiments	Mean S/N ratio
Blank holder force [BHF]	1	1, 2, 3	16.48
	2	4, 5, 6	12.50
	3	7, 8, 9	12.38
Coefficient of friction [μ]	1	1, 4, 7	15.27
	2	2, 5, 8	14.15
	3	3, 6, 9	11.93
Die profile radius [RD]	1	1, 6, 8	14.39
	2	2, 4, 9	14.31
	3	3, 5, 7	12.64
Punch nose radius [RP]	1	1, 5, 9	14.48
	2	2, 6, 7	14.36
	3	3, 4, 8	12.50

Table 1.9 Connector—ANNOVA results for thinning.

	BHF	Friction	RD	RP
1	16.48	15.27	14.39	14.48
2	12.50	14.15	14.31	14.36
3	12.38	11.93	12.64	12.50
Range	4.09	3.33	1.75	1.98
Rank	1	2	4	3

Table 1.10 Connector—S/N ratios in nine experiments for wrinkling.

Expt. no.	Distance square of all wrinkling points from wrinkling curve (mm^2)	S/N ratio
1	1291.27	−62.22
2	1146.14	−61.18
3	1778.80	−65.00
4	1000.83	−60.00
5	994.69	−59.95
6	859.01	−58.67
7	1307.52	−62.32
8	918.74	−59.26
9	1046.89	−60.39

1.11 Analysis of variance for wrinkling—Connector

The wrinkling phenomenon at various cross sections during all nine experiments is presented in Table 1.10. The wrinkling effect is calculated from failure limit diagrams. The addition of the square of vertical distance from the wrinkling curve of all points below the wrinkling curve is presented. The S/N ratios are calculated for the wrinkling quality characteristic of "smaller is better."

The maximum distance of wrinkling points observed from the failure limit diagram is 1778.80 mm^2 and the minimum is 859.01 mm^2. The maximum wrinkling distance is observed in Experiment 3. The analysis of variance results are presented in Tables 1.11 and 1.12. The mean S/N ratios were calculated for all parameters blank holder force, coefficient of friction,

Table 1.11 Connector—S/N ratios at three levels for wrinkling.

Parameter	Level	Experiments	Mean S/N ratio
Blank holder force [BHF]	1	1, 2, 3	−63.42
	2	4, 5, 6	−63.88
	3	7, 8, 9	−64.60
Coefficient of friction [μ]	1	1, 4, 7	−64.90
	2	2, 5, 8	−64.00
	3	3, 6, 9	−63.00
Die profile radius [RD]	1	1, 6, 8	−62.59
	2	2, 4, 9	−64.38
	3	3, 5, 7	−64.93
Punch nose radius [RP]	1	1, 5, 9	−62.79
	2	2, 6, 7	−63.06
	3	3, 4, 8	−66.05

Table 1.12 Connector—ANNOVA results for wrinkling.

	BHF	Friction	RD	RP
1	−63.42	−64.90	−62.59	−62.79
2	−63.88	−64.00	−64.38	−63.06
3	−64.60	−63.00	−64.93	−66.05
Range	0.45	1.90	2.34	3.26
Rank	4	3	2	1

die profile radius and punch nose radius at all three levels of low, medium, and high. The range is defined as the difference between maximum and minimum values of the S/N ratio for the particular parameter.

Table 1.12 shows the rearrangement of S/N ratios for all variables for all levels. The rank indicates the influence of input parameter on the quality characteristic. The result for the preceding orthogonal array indicates that punch nose radius has a major influence on wrinkling. The die profile radius takes second place; friction has third place, and blank holder force has the least influence on wrinkling.

1.12 Linear regression analysis

Regression analysis is a statistical tool for the investigation of relationships between variables. Usually, the investigator seeks to ascertain the causal effect of one variable on another. To explore such issues, the investigator assembles data on the underlying variables of interest and employs regression to estimate the quantitative effect of the causal variables upon the variable that they influence (Naceur et al., 2001; Naceur, Guo, & Batoz, 2004). The investigator also typically assesses the statistical significance of the estimated relationships, that is, the degree of confidence that the true relationship is close to the estimated relationship. At the outset of any regression study, one formulates some hypothesis about the relationship between the variables of interest that are input to the system and some performance characteristics of the system that are dependent on these variables (Gan & Wagoner, 2004; Zhang, Wang, Wang, Xu, & Chen, 2004).

1.13 Mathematical modeling—Connector

Linear mathematical relations have been developed from the results of the Taguchi design of experiments and analysis of variance between the input parameters of blank holder force, friction coefficient, die profile radius, and punch nose radius. The performance characteristics considered for the connector in this research are Thinning and Wrinkling. The relationships are as follows. Minitab has been used for the regression analysis.

$$\text{THINNING} = 1.35 - 0.0400 \, \boldsymbol{BHF} - 0.733 \, \boldsymbol{\mu} - 0.0300 \, \boldsymbol{R_D}$$
$$- 0.0183 \, \boldsymbol{R_P} \tag{1.1}$$

$$\text{WRINKLING} = 349 - 157 \, \boldsymbol{BHF} + 283 \, \boldsymbol{\mu} + 337 \, \boldsymbol{R_D} + 61 \, \boldsymbol{R_P} \tag{1.2}$$

1.14 Problem formulation—Thinning and wrinkling

The multiobjective optimization problem has been formulated from the linear mathematical models developed. Two objectives, thinning and wrinkling, were selected. The formulated problem is as follows:

$$\text{Minimize } \boldsymbol{F} = (\boldsymbol{F1}, \boldsymbol{F2})$$

$$\boldsymbol{F1} = 1.35 - 0.0400 \, \boldsymbol{BHF} - 0.733 \, \boldsymbol{\mu} - 0.0300 \, \boldsymbol{R_D} - 0.0183 \, \boldsymbol{R_P} \tag{1.3}$$

$$\boldsymbol{F2} = 349 - 157 \, \boldsymbol{BHF} + 283 \, \boldsymbol{\mu} + 337 \, \boldsymbol{R_D} + 61 \, \boldsymbol{R_P} \tag{1.4}$$

Subjected to $1.2 \leq \boldsymbol{\beta} \leq 2.2$

$$3 \, \boldsymbol{R_D} \leq \boldsymbol{R_P} \leq 6 \, \boldsymbol{R_D}$$

$$\boldsymbol{F_{d \, Max}} \leq \boldsymbol{\pi} \, \boldsymbol{d_m} \, \boldsymbol{S_0} \, \boldsymbol{S_u} \tag{1.5}$$

$$\boldsymbol{R_D} \geq 0.035 \left[50 + (\boldsymbol{d_0} - \boldsymbol{d_1}) \right] \sqrt{\boldsymbol{S_0}} \tag{1.6}$$

1.15 Multiobjective Multiverse optimization algorithm (MOMVO)

The multiverse optimization algorithm is an evolutionary algorithm that mimics the incidence held in creating the universe, i.e., the theory put forth by scientists and physicists popularly known as fhe Big Bang theory. Based on this theory, the universe came into existence suddenly, and parallel to this theory a new theory came into consideration, which is the multiverse concept. The multiverse concept is that there were many big bangs, with each big bang giving birth to an individual universe, resulting in a multiverse. The multiverse algorithm works off of this basic theory. There are three important concepts used in this algorithm from the multiverse theory: white hole, black hole, and wormhole. The white hole is considered to be a hole present in any universe that tends to throw objects out of it, which can be referred to as a big bang incident. The collisions of one universe with another in a multiverse lead to the generation of a white hole. The black hole is a hole that is considered as opposite to a white hole: it can pull anything into it, including light, due to its extreme gravitational pull. Wormholes are holes that can transfer the objects in the universe from one corner point to another in an instant: they act like a tunnel for passing objects. Every universe in the multiverse is continuously expanding at a fixed rate, which is known as the inflation rate. The objects in it, such as planets, stars, black holes, wormholes, and white holes, are moving (changing their position) at some rate. It is considered that each and every universe in the multiverse is reacting with the others through black holes, white holes, and wormholes. This led to the inspiration for the multiverse algorithm (Mirjalili, Mirjalili, & Hatamlou, 2016).

The algorithm works in two phases: exploration and exploitation. In the multiverse optimization (MVO) algorithm, the black holes and white holes work as a medium for exploration. The wormholes help the algorithm in exploitation, so both exploration and exploitation are achieved in this algorithm. The analogy is that the solution or fitness function value is assumed to be the universe and the objects present in the universe are the variables. In addition to this concept, each and every universe has its rate of inflation; similarly this rate of inflation is provided to each solution in the multiverse (Chen, Li, & Kuang, 2021). For optimization, there are some rules to be followed, as follows:

- With a higher rate of inflation, the probability of having white holes is high.
- With a higher rate of inflation, the probability of having black holes is low.
- A higher probability of white holes present in the universe tends to send objects through the white holes.
- Having a higher probability of black holes present in the universe tends to accept or receive objects through the black holes.
- The objects randomly continue moving through the universe, through the wormholes, regardless of the rate of inflation, until the best universe is found.

the Pseudocode of MVO is shown in Fig. 1.15.

There is always an exchange of objects in a universe in the multiverse system through the white holes and black holes. The universes with a higher inflation rate have a greater number of white holes and they tend to send objects, and the universes with a lower inflation rate have more black holes, which tend to receive the objects (Benmessahel, Xie, & Chellal, 2020). This

```
SU=Sorted universes
NI=Normalize inflation rate (fitness) of the universes
for each universe indexed by i
Black_hole_index=i;
for each object indexed by j
r1=random([0,1]);
if r1<NI(Ui)
White_hole_index= RouletteWheelSelection(-NI);
U (Black_hole_index, j) = SU (White_hole_index, j);
end if
end for
end for
```

Fig. 1.15 Pseudocode of MVO. *No permission required.*

exchange of objects is through tunnels, which are selected by a randomized roulette wheel selection method. The position of an object is calculated by the formulation:

$$
x_i^j = \begin{cases} \begin{cases} X_j + TDR \times \left((ub_j - lb_j) \times r4 + lb_j \right) & r3 < 0.5 \\ X_j - TDR \times \left((ub_j - lb_j) \times r4 + lb_j \right) & r3 \geq 0.5 \end{cases} & r2 < WEP \\ x_i^j & r2 \geq WEP \end{cases}
$$

$r2, r3, r4$ are the random numbers in $[0, 1]$. TDR is traveling distance rate and *WEP* is wormhole existence probability X indicates the best universe found so far. The *ub* and *lb* are the upper bound and lower bound of the variables.

Multiobjective also runs on the same theory, but due to the presence of many best solutions to each objective function, the white hole and wormhole are chosen very particularly. In order to run this algorithm for the multiobjective, some principles are defined as follows:

- If any new solution dominates the previous solution, it should directly replace the previous solution.
- If the new solution does not dominate the previous solution, then it should be removed and never considered.

- If the new solution is nondominated with respect to the present solution, then it should be stored.
- If the decided storage is full, then the unwanted solution should be discarded and the nondominated solution should be stored.

This is the process by which the multiobjective multiverse (MOMVO) is applied and a Pareto graph is plotted with the optimum solution.

1.16 Multiobjective optimization

A multiobjective optimization problem involves a number of objective functions that are to be either minimized or maximized (Gantar, Kuzman, & Filipič, 2005). As in a single-objective optimization problem, the multiobjective optimization problem may contain a number of constraints that any feasible solution (including all optimal solutions) must satisfy. Matlab is used as a tool for optimization. The parameters of a multiobjective problem are presented in Table 1.13.

The results obtained are presented in Table 1.14. One of the optimum values of die profile radius is 2.5 mm and the coefficient of friction is 0.05. The wrinkling parameter is 1220 mm^2 and thinning is observed to be 0.02 mm, meaning the optimized thickness is 0.98 mm.

The Pareto front indicates various combinations of conflicting objectives. For the two objective functions thinning and wrinkling, the Pareto front is as shown in Fig. 1.16.

Table 1.13 Connector—Parameters—Thinning and wrinkling.
MOGA optimization parameters

Population	Double vector
Selection	Tournament
Crossover	Two point
Mutation	Constraint dependent
Migration	Forward
Crossover probability	0.80
Pareto front fraction	0.65
Stopping criterion	Number of generations
Generations	800
Initial population	600

Table 1.14 Connector—MOGA Parameters—Thinning and wrinkling.
MOGA results

Parameter	Lower bound	Upper bound	Optimum
Die profile radius	2.5 mm	3.5 mm	2.5 mm
Friction	0.05	0.15	0.05
Wrinkling parameter		1220 mm^2	
Thinning		0.02 mm	

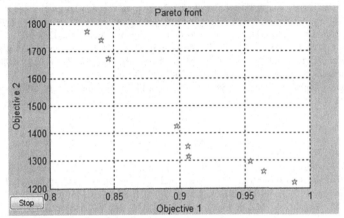

Fig. 1.16 Pareto front—Connector—Thinning and wrinkling. *No permission required.*

1.17 Conclusion

It is obvious from the results that the thinning has been minimized and the achieved thickness is 0.98 mm after optimization. Also, the wrinkling factor is 1220, indicating that a few of the wrinkling points have been shifted to the safe zone, making the connector design and process optimum and safe. The multiverse optimization algorithm performs very well and achieves optimized results.

References

Ahmetoglu, M., Broek, T. R., Kinzel, G., & Altan, T. (1995). Control of blank holder force to eliminate wrinkling and fracture in deep-drawing rectangular parts. *CIRP Annals - Manufacturing Technology, 44*(1), 247–250. https://doi.org/10.1016/S0007-8506(07)62318-X.

Ayed, D., & Batoz, J.-L. (2005). Optimization of the blank holder force with application to Numisheet'99 front door panel. In *VIII International Conference on Computational Plasticity* (pp. 1–4).

Benmessahel, I., Xie, K., & Chellal, M. (2020). A new competitive multiverse optimization technique for solving single-objective and multiobjective problems. *Engineering Reports, 2*(3). https://doi.org/10.1002/eng2.12124.

Bleck, W., Deng, Z., Papamantellos, K., & Gusek, C. O. (1998). A comparative study of the forming-limit diagram models for sheet steels. *Journal of Materials Processing Technology, 83* (1–3), 223–230. https://doi.org/10.1016/s0924-0136(98)00066-1.

Chan, W. M., Chew, H. I., Lee, H. P., & Cheok, B. T. (2004). Finite element analysis of spring-back of V-bending sheet metal forming processes. *Journal of Materials Processing Technology, 148*(1), 15–24. https://doi.org/10.1016/j.jmatprotec.2003.11.038.

Chen, L., Li, L., & Kuang, W. (2021). A hybrid multiverse optimisation algorithm based on differential evolution and adaptive mutation. *Journal of Experimental & Theoretical Artificial Intelligence, 33*(2), 239–261. https://doi.org/10.1080/0952813X.2020.1735532.

Gan, W., & Wagoner, R. H. (2004). Die design method for sheet springback. *International Journal of Mechanical Sciences, 46*(7), 1097–1113. https://doi.org/10.1016/j.ijmecsci.2004.06.006.

Gantar, G., Kuzman, K., & Filipič, B. (2005). Increasing the stability of the deep drawing process by simulation-based optimization. *Journal of Materials Processing Technology, 164–165*, 1343–1350. https://doi.org/10.1016/j.jmatprotec.2005.02.099.

Ghoo, B. Y., & Keum, Y. T. (2000). Expert drawbead models for sectional FEM analysis of sheet metal forming processes. *Journal of Materials Processing Technology, 105*(1), 7–16. https://doi.org/10.1016/S0924-0136(00)00646-4.

Kawka, M., Olejnik, L., Rosochowski, A., Sunaga, H., & Makinouchi, A. (2001). Simulation of wrinkling in sheet metal forming. *Journal of Materials Processing Technology, 109*(3), 283–289. https://doi.org/10.1016/S0924-0136(00)00813-X.

Kim, Y., & Son, Y. (2000). Study on wrinkling limit diagram of anisotropic sheet metals. *Journal of Materials Processing Technology, 97*(1–3), 88–94. https://doi.org/10.1016/s0924-0136(99)00346-5.

Lei, L.-P., Hwang, S.-M., & Kang, B.-S. (2001). Finite element analysis and design in stainless steel sheet forming and its experimental comparison. *Journal of Materials Processing Technology, 110*(1), 70–77. https://doi.org/10.1016/s0924-0136(00)00735-4.

Mirjalili, S., Mirjalili, S. M., & Hatamlou, A. (2016). Multi-verse optimizer: A nature-inspired algorithm for global optimization. *Neural Computing and Applications, 27*(2), 495–513. https://doi.org/10.1007/s00521-015-1870-7.

Moshksar, M. M., & Mansorzadeh, S. (2003). Determination of the forming limit diagram for Al 3105 sheet. *Journal of Materials Processing Technology, 141*(1), 138–142. https://doi.org/10.1016/S0924-0136(03)00262-0.

Naceur, H., Guo, Y. Q., & Batoz, J. L. (2004). Blank optimization in sheet metal forming using an evolutionary algorithm. *Journal of Materials Processing Technology, 151*(1–3), 183–191. https://doi.org/10.1016/j.jmatprotec.2004.04.036.

Naceur, H., Guo, Y. Q., Batoz, J. L., & Knopf-Lenoir, C. (2001). Optimization of drawbead restraining forces and drawbead design in sheet metal forming process. *International Journal of Mechanical Sciences, 43*(10), 2407–2434. https://doi.org/10.1016/S0020-7403(01)00014-5.

Obermeyer, E. J., & Majlessi, S. A. (1998). A review of recent advances in the application of blank-holder force towards improving the forming limits of sheet metal parts. *Journal of Materials Processing Technology, 75*(1–3), 222–234. https://doi.org/10.1016/S0924-0136(97)00368-3.

Ozturk, F., & Lee, D. (2004). Analysis of forming limits using ductile fracture criteria. *Journal of Materials Processing Technology, 147*(3), 397–404. https://doi.org/10.1016/j.jmatprotec.2004.01.014.

Xia, Q., Shima, S., Kotera, H., & Yasuhuku, D. (2005). A study of the one-path deep drawing spinning of cups. *Journal of Materials Processing Technology, 159*(3), 397–400. https://doi.org/10.1016/j.jmatprotec.2004.05.027.

Zhang, S. H., Wang, Z. R., Wang, Z. T., Xu, Y., & Chen, K. B. (2004). Some new features in the development of metal forming technology. *Journal of Materials Processing Technology, 151*(1–3), 39–47. https://doi.org/10.1016/j.jmatprotec.2004.04.098.

Zimniak, Z. (2000). Problems of multi-step forming sheet metal process design. *Journal of Materials Processing Technology, 106*(1–3), 152–158. https://doi.org/10.1016/S0924-0136(00)00607-5.

CHAPTER TWO

An approach for machining curve cooling hole in plastic injection mold

Diwesh Babruwan Meshram[a], Yogesh M. Puri[b], Sachin Ambade[c], Vikas Gohil[d], and Ganadhar Rajaram Navnage[e]

[a]Department of Plastics Engineering, Central Institute of Petrochemicals Engineering and Technology, Korba, India
[b]Department of Mechanical Engineering, VNIT, Nagpur, India
[c]Department of Mechanical Engineering, Yeshwantrao Chavan College of Engineering, Nagpur, India
[d]Mechanical Engineering Department, Bajaj Institute of Technology, Pipri, Wardha, India
[e]Department of Metallurgical Engineering and Materials Science, VNIT, Nagpur, India

2.1 Introduction

Electrical discharge machining (*EDM*) is an advanced and non-conventional machining mechanism to machine the *workpiece* through the tool by spark erosion principle. Its application frequency is very high in the die and mold industry. Various other applications include aerospace, medical, marine, and automobile. The die-sinking *EDM* is the frequently applied mechanism for such applications. The *curvilinear* motion of the tool is the key control approach for machining the *workpiece*. The optimal setting of the independent input machining parameters is an important process for the expected outcome. The structure of the proposed curved hole in the *workpiece* lies with the differential trajectory. Thus, a new programmable electronic control system (*ECS*) is designed and developed to synchronize the required output motion. Machining the curved hole by employing a novel programmable *ECS* and estimating the quality of execution based on experimental validation is the key objective of this chapter. It interacts with the forward-reverse motion of curved tool on stationary *workpiece,* which results in the generation of spark. A new way of machining a curved cooling hole through a curved tool is possible by z-axis numerical control (*ZNC*) *EDM*. Advanced machining or non-conventional machining has been applied in modern machining since 1943 when *Lazarenko* experimented with erosion by electrical sparking. They studied erosion by the theory of electrical discharging. The control of erosion is possible

Computational Intelligence in Manufacturing
https://doi.org/10.1016/B978-0-323-91854-1.00005-4

if the tool is immersed in the dielectric fluid. The developed compact *EDM* mechanism performs the machining internally and externally on the high-carbon high-chromium (*HCHCR*) die steel, oil-hardening non-shrinkage steel (*OHNS*), and *P20* mold steel *workpiece* material, respectively.

Initially, the *ZNC EDM* machine utilizes a copper tool machined on the tool steel. The precision machining is achieved by setting the machining parameters, i.e., spark current and pulse on time (Ramuvel & Paramasivam, 2020). However, Nair, Dutta, Narayanan, and Giridharan (2019) presented an excellent *EDM* process applied on *Ti6Al4V workpiece* material and brass tool. Its *electrochemical* characteristics are analyzed in his research work. As a result, a negative polarity brass tool is applied to perform the machining. There are two key domains, discharge energy and latent heat of vaporization of *workpiece* material selected to investigate the experimental characteristics. The confirmation runs were performed with a 16.89% error over the diameter of the crater. This is generalized in the present chapter with distinct materials of the *workpiece* and tool. Recently, a key development in *EDM* is based on *graphene nanofluid*. Paswan, Pramanik, and Chattopadhyaya (2020) applied this mechanism over *Inconel* 718, and correspondingly, an improved performance is achieved in this chapter. The field emission scanning electron microscopy (FE-*SEM*) image-based analysis is examined in this chapter.

Further, Wang, Yi, Easton, and Ding (2020) proposed the *EDM* process to machine the poly-crystalline diamond *workpiece* material. An advanced *EDM* has appeared with the improvement of 10 times higher than the existed *EDM* as well as a 70% reduced tool wear rate (*TWR*) compared to the equivalent *EDM* process. The proposed chapter achieved an improved performance as 66.32 mm total machining depth by 480 min of machining time. In 2015, the author performed the *EDM* process over the distinct tools (D'Urso, Maccarini, Quarto, & Ravasio, 2015). Micro-*EDM* stainless steel is applied in his experiment. Next year, its extension appears. The author performed this experiment adjoined with a copper tool (D'Urso, Maccarini, Quarto, Ravasio, & Caldara, 2016). The variants of tools and their structure are analyzed in this application. In 2015, another noteworthy application existed on the rate of material removal. *Bahgat* studied the constant *EDM* process (Bahgat, Shash, Abd-Rabou, & El-Mahallawi, 2019).

In the die-sinking *EDM* process, the linear hole is developed through the cylindrical tool attached to the *EDM* head, which is reciprocating in the vertical axis on the *workpiece* with a certain spark gap. Due to certain limitations of the *EDM* head and axis, the size and depth of the developed hole are limited. Although the alternative machining methods have been established and implemented for the development of the hole, they had some specific

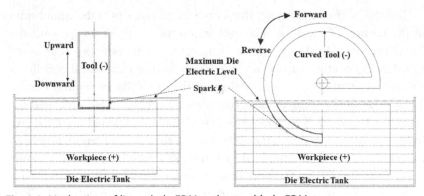

Fig. 2.1 Mechanism of linear hole EDM and curved hole EDM.

limitations, which restrict to adapt the machining process. With the recent development in *EDM*, the customized requirements in the industries are increasing. To overcome restrictions, it is necessary to evaluate an alternative mechanism that can easily machine the curvature hole in two axes.

The author proposed the theory of material removal rate (*MRR*) and tool wear rate (*TWR*) over the copper tool for the distinct *EDM* process (Straka & Hašová, 2018). This is referred to as the die-sinking process over the tool steel. The mechanism of linear hole *EDM* and curved hole *EDM* is presented in Fig. 2.1.

The curved hole *EDM* principle involves the curved tool rotating through a specific direction of the axis of rotation with a certain angular speed. The forward and reverse movement of the curved tool provides the curvature direction and in-depth penetration during machining on the *workpiece*.

In 2004, a long curved hole was developed by EDM based on in-pipe movable mechanism. This opens a new dimension of the EDM process through variable parameters (Ishida, Kogure, Miyake, & Takeuchi, 2004). The author applied the mole-based EDM in 1989 (Fukui & Kinoshita, 1989). In 1993, another variant of the EDM process as a curved passage has designed (Guirguis & Kamal, 1993). Ishida used automation in EDM in 1999. An automatic discharge gap controller is applied in this new mechanism (Ishida & Takeuchi, 1999). Recently, a new approach has been experimented as a suspended ball mechanism adjoined with the conventional EDM process (Yamaguchi, Okada, & Miyake, 2015). In 2017, the author improved the preceding EDM process (Okada, Yamaguchi, & Ota, 2017). Another geometrical structure came into existence in 2002. The curved motion generator designs the L-shaped curved hole. Ishida generalized the EDM by this new variant inspired by geometry (Ishida & Takeuchi, 2002b).

In 2005, constant curvature theory was in existence over the equilibrium of the motion of the tool. The author introduced this concept with the *EDM* process (Miyake, Ishida, & Takeuchi, 2005). *Bayramoglu* analyzed *EDM* by the contour theory in 1995. Contour-based linear and non-linear structure is adopted and processed through the *EDM* (Bayramoglu & Duffill, 1995). They applied the circular contour for the same. *Ishida* operated the tool over the slider chain cranks in 2002 (Ishida & Takeuchi, 2002a). An improvement is recorded in the *EDM* process than the conventional process by implementing the customized tool through a curved structure. The survey of the proposed work is presented in a lucid way (Slatineanu et al., 2011) in the same year. Further, the author analyzed the curved hole *EDM* machining by the discrete parameters in 2004 (Uchiyama & Shibazaki, 2004). In the same year, *Ishida* extended the result by a long curved mechanism using the in-pipe system (Ishida et al., 2004). *Ishida* presented the extension of the preceding work in 2005 (Ishida, Kogure, Miyake, & Takeuchi, 2005). However, *Ishida* established this idea in the year 2005 (Ishida, Nakajima, Miyake, & Takeuchi, 2005). Next year, *Nakajima* applied optimization for reducing the diameter of the hole (Nakajima, Ishida, Kita, Teramoto, & Takeuchi, 2006). In 2009, the development of the *EDM* process was published by Mot (Mot & Purcar, 2009).

Uliuliuc designed *curvilinear* axis holes for the *EDM* process in 2010 (Uliuliuc, Antonio, Coelho, & Slatineanuc, 2010). This mechanism improves the process by the continuous distribution of discharging. *Ishida* proposed a similar mechanism in 2008 (Ishida, Nagasawa, Kita, & Takeuchi, n.d.). They applied the hemisphere instead of the *curvilinear* structure for the *EDM* system. *Kita* generalized the *EDM* process by the reduction of the size of the gap between the *workpiece* and tool (Kita, Ishida, Teramoto, & Takeuchi, 2010). This improves the performance of the process. *Ishida,* in 2010, improved the preceding mechanism (Ishida & Takeuchi, 2010). The optimized gap controller system is applied in the proposed mechanism.

In 2009, a cross section-based *EDM* process was introduced by *Ishida* (Ishida, Mochizuki, & Takeuchi, 2009) for *EDM* processing under the context of variable input operation. *Ishida* discussed an efficient *EDM* process by applying curvature methodology and optimized control-based trajectory (Ishida et al., 2014; Ishida & Takeuchi, 2008). In 2013, *Ishiguro* formulated the *EDM* process over the CAD/CAM application (Ishiguro, Ishida, Kita, Nakamoto, & Takeuchi, 2009). The *literatures* (Ids & Clearfield, 1998; Karunakaran, 2010) covered these applied methodologies. Recently, *Pushyanath* presented the new mechanism on *HCHCR AISI-D7*-based

EDM process (Pushyanth & Bhaskar, 2018). In 2018, another variant of the EDM process came into existence (Praveen, Geeta Krishna, Venugopal, & Prasad, 2018). The application of pulse-on time and pulse-off time is transformed into the said system for the EDM process. The classification of the set-based EDM system is introduced by Lin in 2002 (Lin, Lin, & Ko, 2002). Fuzzy-based EDM is presented to process the machining controlled by gray relational analysis (GRA). This system improves the whole EDM process simultaneously.

To meet the requirements, the author has proposed a new method of machining even the compound curvature of the hole using simple additional equipment. The advantages of this additional equipment involve easy to manufacture, two axes machining, and internal and external hole development. The specific application of the proposed setup includes the development of a curved cooling channel for the plastic injection mold and surgical instrument for medical application. Its form is not limited to the above but can be used in automobile and aerospace applications. Recently, an advanced EDM mechanism is developed based on rotary tool near-dry characteristics (Yadav, Kumar, & Dvivedi, 2019). There are high-speed steel (HSS) materials of the workpiece and the copper tool used in the EDM process. This EDM process performs the machining in the oxygen gas-based dielectric medium. The rate of supply of oxygen gas is the key factor to perform the machining. The proposed chapter used spark erosion oil as die-sinking EDM. There is a 203% improvement in MRR with dry EDM using air, but it reduces to 110% with dry air as oxygen. In the same year, the author performed the curved hole machining of the workpiece based on the OHNS material through the semicircular curved copper tool (Meshram & Puri, 2019). It machined a curvature channel. Its effectiveness is proved by modern analytic tools, i.e., ANOVA, Taguchi, regression analysis, and its compact method over ECS. An optimized result is obtained under the context of optimum MRR and EWR. The rate of MRR and EWR lies with the level of significance. The semicircular structure with different nodes proves the concept of geometry. The confirmation runs hold the four combinations of the machining parameters that correspond with the four decision variables. This chapter proposes a discrete dimension of the EDM through the structural and electrochemical characteristics of the tool and workpiece simultaneously. Another EDM mechanism came into existence in 2020. The author discussed the concept of machining linear and non-linear holes simultaneously (Meshram & Puri, 2020). This is also based on the optimized analysis of the existed EDM mechanism. The compact optimized technique of ANOVA, Taguchi, TOPSIS, and statistical analysis is applied for the developed *EDM*. Basically, this result occurred through the analysis of the oscillation motion of the tool. There is minimized error by the said optimized techniques. Thus, the proposed *EDM* is used for machining conventional and non-conventional both types of holes efficiently. This puts a foundation to cut chaotically through the advanced *EDM* over the random environment.

2.2 Methodology

Fig. 2.2 illustrates the isometric view with the experimental flow of the device proposed in this work. The individual components are designed and assembled using the Siemens (*NX* 11.0) software. The selection of components based on mechanical and electrical controls is validated by performing the planned experiments. A programmable *ECS* is implemented for synchronizing and achieving the curved hole machining. The curved copper tool with three categories of *workpiece* material is utilized for the validation of the proposed mechanism. Initially, a rectangular cross section curved copper tool with *HCHCR workpiece* is used for the pilot experimentation. Further, the performance evaluation is conducted by executing the two combinations of the curved tool and *workpiece* material. Customized solid and hollow curved copper tool (cross section: circular) is used with *OHNS* and *P20 workpiece* material, respectively (primary and secondary experimentation). The complete operating procedure of the proposed

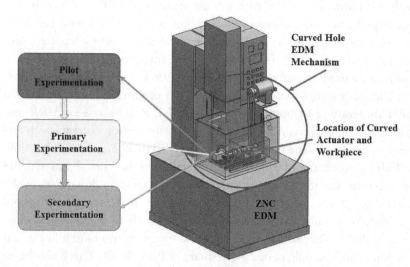

Fig. 2.2 Schematic assembled view of the proposed device with experimental methodology.

mechanism and the selection of machining parameters with planned experiments are the two critical areas in generating the curved cooling hole using
ZNC EDM.

2.2.1 Mechanical–electrical methodology

The methodology is based on the compact theory of physical and chemical
composition. All the components are placed inside the working tank. The significant components are permanent magnet direct control (*PMDC*) servomotor with a *tacho*generator and curved tool. The complete transmission of the
motion through both components relies on the outputs. The oscillating
motion is transmitted from the *PMDC* servomotor to the small aluminum
timing gear ($\varnothing 8.42\,mm$, width—*15mm*, teeth—19 numbers) and big aluminum timing gear ($\varnothing$ *94.42mm*, width—*15mm*, teeth—100 numbers) via
PowerGrip *HTD* timing belt (total length—*753mm*, width—*15mm*,
pitch—*3mm*, tooth height—1.*2mm*). Two L-shaped mild steel jigs
(*260mm* × *152mm* × *300mm*, *178mm* × *152mm* × *154mm*) with thickness
8mm are used for holding the *PMDC* servomotor. They also play an important role in aligning and adjusting the timing gears and belts. Though the transmission speed is very less, a bearing support assembly (Model: *SYJ 25TF-SKF*)
is implemented. This assembly is rested on the six hollow aluminum rectangular channels. The channels are locked with four nylon plastic hexagonal
bolts with nuts (*M10* x 145mm). It acts as the insulating device between
the working tank and the installed device components. Most components
are designed and selected based on the concept of weight reduction. Further,
a lightweight hollow acrylic transparent box (200mm × 160mm × 160mm,
thickness—*5mm*) is provided inside the working tank to prevent the components from rusting and other environmental issues. The connection shaft is
provided inside the bearing support assembly. The endpoints of the shaft
are coupled with the big timing gear and *collet* arrangement. The collet
(*ER16,∅ 10mm*) is coupled with hexagonal nut and bolt, which act as a
locking mechanism for the curved tool. Similarly, a novel tool holding mechanism (*THM*) is designed and developed for holding and aligning the curved
tool. The necessary current required for performing the curve hole machining
is provided by the power cable with *multicore*-coated wire of $\varnothing 3.25\,mm$. Its
maximum current holding capacity is *20A*. Additionally, a programmable
ECS is placed on the right side of the machine. The desired oscillating motion
is obtained by the optical sensor located at the upper side of the ram head. This
sensor transmits the signals received from the linear scale provided at ram head
to the programmable *ECS*. The programmable *ECS* commands the *PMDC*

servomotor for the desired motion. The curved tool and its desired motion hold a crucial role in the proposed mechanism. Fig. 2.3 depicts the front and side schematic views of a mechanism installed on the *ZNC EDM* machine.

2.2.2 Programmable ECS

The ECS used in this study is Arduino (Model: Mega 2560), which acts as the main microcontroller. It will operate according to the prewritten instruction into the code memory of Arduino. Optocoupler (Model: B817) is used to isolate the power supply between two circuits that include Arduino mega circuit and motor driver circuit. It also removes the noise and transient voltages and prevents latch-up or malfunctioning. The signal is transferred optically from Arduino mega to H-bridge driver. It is used to amplify the signal from the microcontroller in the required level to the motor and also its voltage and current. In H-bridge, four power transistors (two transistor Models:- TIP147, two transistor Models:- TIP142) are joined, and it allows the speed control with the bidirectional motor. The linear power supply is used to obtain the required voltage (34 V) and current (5A) using a step-down transformer, bridge rectifier, and a capacitor. More-over, H-bridge driver has been interlinked between DC power supply and DC servomotor. This power supply has been connected to the Arduino and Arduino (Model: Uno R3). Next, the adapter is connected through the small linear power supply. Optical sensor and graphical LCD are the additional accessories used in this device. The microcontroller read the command from the membrane, i.e., keypad (4 × 4) as the input. The output device used to display the command with the keypad number is the graphical LCD (128-pixel x 64-pixel). Arduino reset circuit is used in this device to reset the microcontroller at a specific interval of time. Due to the non-availability of the inbuilt reset circuit in Arduino Mega, Arduino reset circuit can latch-up up to 500 ms for the average working. A toggle button is used to prevent reset during the reprogramming of Arduino mega and placed between the Arduino reset circuit and Arduino mega. Bidirectional data communication has been attached through the single wire communication to the controller. Additionally, the voltage sensing circuit is used to convert the high AC voltage into low DC voltage improved using inbuilt ADC in Arduino. It senses the motor speed of EDM using an analog tachometer attached to EDM motor. The circuit diagram of the ECS is shown in Fig. 2.4.

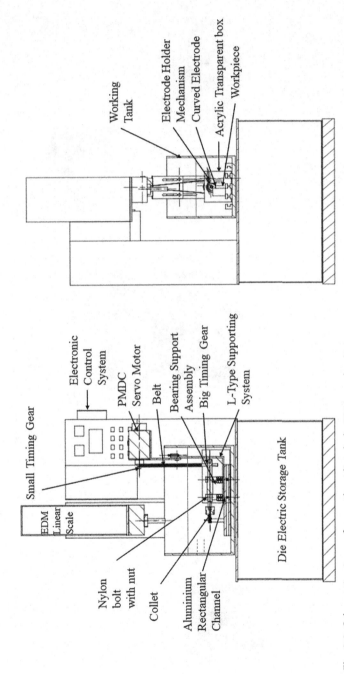

Fig. 2.3 Schematic views of curved cooling hole EDM mechanism.

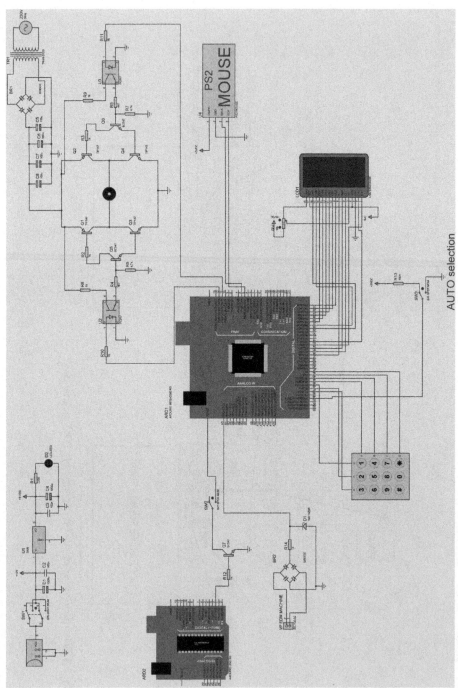

AUTO selection

Fig. 2.4 Electronic control system.

2.2.3 Curved tool and workpiece material

Based on the selection of mechanical, electrical, and ECSs used in this mechanism, a curved tool with different geometries and workpieces are selected. Initially, a curved tool (Ø 70 mm, thickness—10 mm, width—12 mm) and HCHCR workpiece material (127 mm × 60 mm × 16 mm) with a rectangular cross section have been experimented without the use of EHM. Further, the two experiments are performed based on the circular cross section curved tool with EHM. The first experiment utilizes the solid curved copper tool (Ø 32 mm, thickness—3.6 mm) with rectangular OHNS workpiece material (40 mm × 40 mm × 12.5 mm). P20 mold steel workpiece material (64 mm x64 mm x 15 mm) and hollow curved copper tool (Ø 32 mm, thickness—3.6 mm) are implemented in the second experiment. The spark emission spectroscopy is used to identify the chemical composition of the materials as given in Table 2.1.

2.3 Experimentation

The evaluation criterion for any developed mechanism or process is experimentation. The curved cooling hole *EDM* mechanism operation and experimentation are discussed below.

2.3.1 Operation of the proposed mechanism

The innovative mechanism of curved hole machining is useful for the curved cooling generation on any material of conducting nature. The actual operation of the device is defined and its operating procedure prepared. However, the *workpiece* and tool have been placed in their respective location. Initially, the machine needs to be switched "ON," and its basic control systems with the display unit need to be activated. The next step is to provide the power source to the *ECS* and *PMDC* servomotor supply with the LCD

Table 2.1 Chemical composition of the workpiece materials.

Workpiece material	Composition (%)				
	C (carbon)	Mn (magnesium)	W (tungsten)	Cr (chromium)	V (vanadium)
HCHCR die steel	1.62	0.28	0.52	11.28	0.42
OHNS	1	1.41	0.5	0.08	0.2
P20 mold steel	0.32	0.75	0.62	1.75	0.39

screen provided on the side panel of the machine. The unique five keys on
the keypad, i.e., A-Forward motion (Motor), B-Reverse Motion (Motor),
C-*APOS* (Automatic Positioning), D-Stop, and "#" Auto, have been used
for various operation features. Before the machining operation, press "*D*"
key for ensuring the restricted motion between the *PMDC* servomotor
and a linear ram of *EDM* machine. The linear ram of the machine is adjusted
as per the tool profile. The tool is placed at a certain distance from the *workpiece* to perform the automatic positioning system (*APOS*) option in the keypad provided. In this device, the initial step is to touch the tool on the
workpiece surface by pressing a key "*C.*" Next, the *APOS* option has been
pushed in the machine display unit. The tool will move slowly forward
and touches the *workpiece*, which results in the activation of the buzzer.
As soon as the buzzer sound energizes, the STOP button is pressed on
the display unit. The digital readout system (*DRO*) button followed by
the ENTER button has been pushed for the zero position of the tool. Once
the upward button gets pushed, the curved tool will move backward at a
certain distance from the zero position to maintain the spark gap. The effective connectivity between the tool and *workpiece* has been disconnected from
the electronic circuit by pressing button "*D.*" As per the requirement, the
linear ram head is moved upward to the extreme position. After the "*C*" key
is pressed, the next key will be "#" in which the auto mode activates. In auto
mode, minor changes in the linear scale affect the motion feedback system
and rotate the tool at a certain speed simultaneously in forward and reverse
directions. Finally, the spark key is pressed in the display unit of the machine
and there is the initialization of the curved hole machining process.

Further, the experiments are conducted under different machining conditions for machining a curved hole. Fig. 2.5 shows the actual configuration
of the curved tool and *workpiece* attached with a curved hole mechanism.

2.3.2 Pilot experimentation

The proposed mechanism is operated, and the pilot experimentation executes its feasibility test. The combination of the rectangular cross section
curved copper tool with HCHCR workpiece material is machined for
the generation of a curved cooling hole. Selected machining parameters
are identified, and their operating values are fixed as presented in
Table 2.2. The geometry of curved tool is manufactured using the wire
cut EDM. The HCHCR workpiece material is specifically used in the
die and mold application. Its sizing in the required dimension is performed

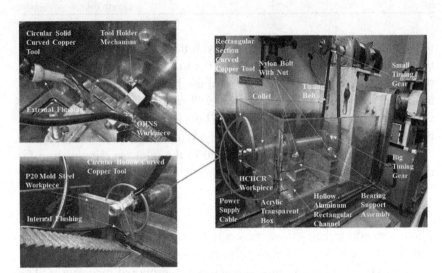

Fig. 2.5 Curved tool and workpiece configuration.

Table 2.2 Experimental conditions.

Sr. no	Selected machining parameters	Operating range	Pilot experiment specification	Primary experiment specification	Secondary experiment specification
1	Tool geometry		Rectangular curved copper	Circular solid curved copper	Circular hollow curved copper
2	Workpiece material		HCHCR	OHNS	P20 mold steel
3	Dielectric medium		Spark Erosion Oil 450		
4	Pulse on time (T-on), μ	(1–99)	25	30	40
5	Duty cycle (τ), μ	(1–99)	9	9	5
6	Sparking current (IP), A	(0–35)	30	10	17
7	Bi-pulse current (IB), A	(0–3)	1	3	2
8	Spark time (SPK), sec	(0 − 20)	11	20	4
9	Lift (LFT), sec	(0.0–2.0)	1	1	2
10	Gap voltage (GAP), V	(01–99)	40	35	50
11	Sensitivity (SEN), mm/min	(01–99)	55	50	50

using a hacksaw machine. Initially, EHM is not used in pilot experimentation. The experiment is executed with the machining time of 480 min. In this pilot experimentation, our aim is to machine the curved cooling hole inside the workpiece. The curved tool will enter from the top face of the workpiece and exit. The curved cooling hole is generated, and it is split into two halves. This is required for identifying the trajectory moved by the curved tool. The natural flushing is used in the pilot experimentation for the eroded material.

2.3.3 Primary experimentation

The proposed mechanism was successfully implemented in machining the curved cooling hole by pilot experimentation. Further, more experiments are required for validating the accuracy of the proposed mechanism. The primary experiment was performed with variable curved tool, workpiece material, and machining parameters. This helps in identifying the variations observed in curved cooling hole machining under defined circumstances. The circular cross section curved tool with OHNS workpiece material sets another dimension for generating the curved hole. It improved the accuracy of machining by implementing the EHM. The EHM is designed and manufactured for the defined radius of the curved tool. It guides the tool in a predefined trajectory with forward-reverse motion. The customized tool is developed on the CNC bending machine with the required curvature radius. OHNS workpiece material is used in this experiment as per the required dimension. This combination is used with external flushing for the removal of excess material from the workpiece. Our aim in this experimentation is to machine the external curved hole on the workpiece. The effect of external machining on internal machined is studied in this primary experimentation. The curved hole is easily visible, and its characteristics are analyzed. The machining time is reduced to 120 min for observing the machining capability.

2.3.4 Secondary experimentation

The secondary experimentation is performed for analyzing the effect of curved hole machining by the circular cross section hollow curved tool on the *P20* mold steel *workpiece* material. The manufacturing process adopted for the curved tool is to bend the hollow tube to the required radius. Further, a *P20* mold steel *workpiece* material is used for generating the curved hole. This material is mostly preferred in the plastic injection mold. The internal flushing mechanism is adopted in a hollow curved tool for material removal during machining. The pressurized dielectric is circulated inside the curved tool during machining. Compared to the external flushing, internal flushing removes the eroded material smoothly and increases the machining efficiency. Further, the machining time is set as 30 min.

The experiments are successfully conducted for machining the curved cooling hole using the developed setup on *ZNC EDM*. The possibility of machining the curvature hole and effectiveness is measured using curved hole *ZNC EDM*. The experimental conditions are given in Table 2.2. The salient features of the curved hole in manufacturing a plastic injection mold include:

- Replacement of the linear cooling hole by a curved cooling hole.
- Increased rate of heat transfer.
- Smooth flow of the coolant inside the curved hole in comparison with conventional drilling.

Hence, the expected result is determined. This is analyzed under the context of recommended and standard parameters of the existed machining described below.

2.4 Results and discussion

In this curved machining device, the cooling hole was easily machined and generated the *curvilinear* path. By using this device, the machining of variable cross section changing slots is possible. Fig. 2.6 is a view of the curved

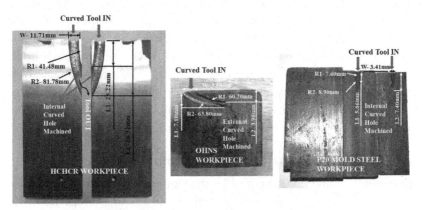

Fig. 2.6 Machined curved hole on three workpiece materials by ZNC EDM.

hole machined by *EDM* with three possible experimentations. The various characterization analyses are discussed below.

2.4.1 Analysis of TWR and MRR

TWR and *MRR* are the major characteristics of the curved hole *EDM*. The removal of material from tool and *workpiece* in a curvature trajectory is analyzed. The governing equation for measuring the *TWR* and *MRR* is given below:

$$\text{TWR}\left(\text{mm}^3/\min\right) = \frac{\text{Tool weight (before)} - \text{Tool weight (after)}}{(\text{Machining time}) \times (\text{Density of Tool})}$$

$$\text{MRR}\left(\text{mm}^3/\min\right) = \frac{\text{Workpiece weight(before)} - \text{Workpiece weight(after)}}{(\text{Machining time}) \times (\text{Density of Workpiece})}$$

where tool weight (before) is the initial weight of the curved tool before machining, tool weight (after) is the final weight of the curved tool after machining, *workpiece* weight (before) is the initial weight of the *workpiece* before machining, *workpiece* weight (after) is the final weight of the *workpiece* after machining, and machining time is the actual machining time required for generating curved geometry.

The measured values of the *TWR* and *MRR* in the pilot experimentation are found to be 4.904 mm³/min and 51.304mm³/min. It covers the complete trajectory profile of the curved hole. Further, the primary experiment is executed with the variation in the machining parameters, curved tool, and *workpiece* material. The reduction in the machining time affects the partial generation of the curved hole. *TWR* and *MRR* values are found to be 1.13 mm³/min and 11.45 mm³/min. Both experiments are performed, and the final validation is through the second experimentation. It is found that 0.3 mm³/min and 2.30 mm³/min are the less values of *TWR* and *MRR*. All the measured values of *TWR* and *MRR* are measured thrice, and the cumulative average values are reported. The experimental errors are also reduced by executing triplicate experiments. From the results of *TWR* and *MRR*, it is observed that the combination of the curved tool and *workpiece* material varies with respect to machining time.

2.4.2 Dimensional analysis of curved hole machined

The curved cooling hole obtained by the various experiments needs to be evaluated for dimensional accuracy with 5 major dimensional factors as mentioned in Table 2.2. The machining *workpiece* is majorly considered for the output result. The entry and exit of the curved tool are defined with the internal and external machining. All the dimensions are measured using coordinate measuring machine (*CMM*). The variation is observed in the arc length due to offset distance provided between the axes of the curved tool and *workpiece*. The comparative analysis of the design values and actual values is analyzed and discussed in Table 2.3. The percentage improvement in the accuracy of pilot, preliminary, and secondary experimentations is 96.87%, 96.20%, and 90.42%, respectively.

2.4.3 Analysis of surface roughness

Surface roughness is another essential and measurable characteristic in the curved hole *EDM*. It is performed by using the surface roughness tester. The results will be used to identify the influence of the machined surface by the curved tool. The measured value (Ra) of curved hole machined on *HCHCR, OHNS,* and *P20* mold steel *workpiece* material is 3.641 μm, 3.242 μm, and 4.609 μm, respectively. The values are measured thrice, and the mean average values for the corresponding measurement were reported. Significant surface variation is observed in curved hole machined in *P20* mold steel *workpiece* material. The remaining two *workpieces* have a sustainable variation in curved hole *EDM*.

2.4.4 Scanning electron microscopy (*SEM*) analysis

The machine surface analysis is performed for the characterization of material under the experimentation methodology using *SEM (JEOL, JCM*-6000 PLUS). The *SEM* analysis is performed on the machined curved hole of *workpiece* materials. This analysis provides the magnified image with the various critical areas formed under specific machining conditions. The *microstructure* analysis on the *workpiece* material is performed before and after the curved hole machining. Fig. 2.7 shows the *HCHCR SEM* image captured at *x75* before and after the machining condition, wherein the flakes are randomly distributed over the machined surface and globules of the smaller and irregular size are identified. The smooth surface with the black dots is the preliminary identifying characteristic before subjecting *workpiece* material to curved hole machining.

Further, the primary experimentation results show that the curved hole is machined on the *OHNS workpiece* material. Fig. 2.8 is the observed surface that consists of the white layers with the porous black holes in before condition. The machined surface is rough, and white flakes are distributed over

Table 2.3 Dimensional analysis of the curved hole machined.

Flow of experiment

	CAD value (mm)	Actual measured value (mm)	Accuracy (%)	CAD value (mm)	Actual measured value (mm)	Accuracy (%)	CAD value (mm)	Actual measured value (mm)	Accuracy (%)	CAD value (mm)	Actual measured value (mm)	Accuracy (%)	CAD value (mm)	Actual measured value (mm)	Accuracy (%)	Average accuracy (%)
Pilot experiment	12	11.71	97.58	30	28.22	94.07	68	66.32	97.53	12	11.71	97.58	12	11.71	97.58	96.87
Primary experiment	3.6	3.58	99.44	8	7.1	88.75	4	3.9	97.5	62	60.24	97.16	65	63.8	98.15	96.2
Secondary experiment	3.6	3.41	94.72	7	5.66	80.86	8	7.4	92.5	8	7.6	95	10	8.9	89	90.42

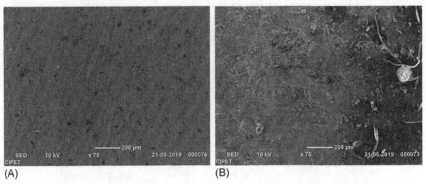

(A)   (B)

Fig. 2.7 SEM image of HCHCR workpiece material. (A) Before machining and (B) After machining.

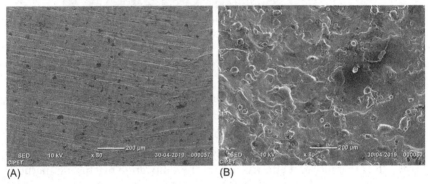

(A)   (B)

Fig. 2.8 SEM image of OHNS workpiece material. (A) Before machining and (B) after machining.

the machined area. Due to the heat generation, the material bonding may be distributed, resulting in the formation of globules. The black–spotted area identifies the primary carbon deposition on the *workpiece* material in after condition. The irregularities in the *workpiece* material affect the properties of the material.

The curved hole machined on the *P20* mold steel *workpiece* material is analyzed for material properties variation. The rough surface consists of white layers with the minor cracks in before machining condition (Fig. 2.9). The small granules are also observed on the rough surface. The major observation is identified in the machined area. The sparking affects the irregular surface distortion in the *workpiece* material. The upper and lower peaks are identified on the surface because of the pulsating spark by a hollow

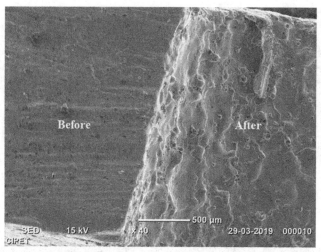

Fig. 2.9 SEM image of P20 mold steel workpiece material showing before and after machining surface morphology.

curved tool. The upper portion shows the L shape area, which indicates that the tool is worn out at the entry of the tool.

All the dimensions related to the curved hole *EDM* are presented. The dimensional variation is observed in the arc length due to the offset distance provided between axes of the curved tool and the *workpiece*. This result confirmed the feasibility and effectiveness of the innovative machining method whereby the tool executes a curved hole on the *workpiece*. A programmable *ECS* over the structure of tool and the material of *workpiece,* respectively, is performed on the above experiments. The experimental methodology set the developed system as an efficient mechanism. The machining time is recorded as a significant coincidence with the existed results. It concludes next.

2.5 Conclusions

The new mechanism of curvature cooling channel machining is proposed in this paper. The maximum value of the *MRR* and the minimum value of the *TWR* are 51.304 mm^3/min (high–carbon high–chromium (*HCHCR*) die steel) and 0.3 mm^3/min (*P20* mold steel), respectively. There is the outcome of this paper as the total machining depth is 66.32 mm in 480 min of machining. This improvement is observed in curved hole machining instead of linear machining. The proposed *EDM* is defined over the curved sparking principle through the copper tool. As a resultant, curved tool material is selected as the copper, and its structure lies with the variation of the motion of the tool forward and backward simultaneously. Thus, a new mechanism of curvature hole machining is proposed in this paper. This is an optimized representation of the *ZNC*-based *EDM* process. The proposed curved hole *EDM* process interacts with the continuous and consistent discharging of the tool and its homogeneous distribution of sparks between the gap of tool and *workpiece*. The dynamic machining in *HCHCR, OHNS,* and *P20 mold* steel *workpiece* through copper curved tool with a varying cross section as non-linear is the key performance of this paper. The experimentation of the proposed mechanism lies with the electrical, electronics, and optimized mechanical-derived compact and feasible design. The experimentation, performance, and comparison are presented in this paper under the context of the existed mechanism. The *EDM* process is compared over the input machining parameters, functional parameters, and output parameters to the equivalent *ZNC* and respective variants of the *EDM* process. There are four outcomes observed in this paper. First, the curved hole is successfully machined in *HCHCR workpiece* material by the copper curved tool. It is achieved by the programmable *ECS* synchronized with all the components in the proposed mechanism. Second, the dimensional study of the curved hole justifies the accuracy of the designed values compared with the actual values. The average accuracy is above 95%, which shows that all the dimensions are accepted for further analysis. The third improvement is noted as surface roughness. The curved hole machined surface may be used as a cooling channel for circulating the coolant in the plastic injection mold. The last advantage is the detailed analysis of the *workpiece* material by the *SEM* technique. The material morphology of the curved hole machined provides a unique solution for the *workpiece* material selection. Finally, the implementation cost of the curvature hole-generated mechanism is estimated for industrial applications.

References

Bahgat, M. M., Shash, A. Y., Abd-Rabou, M., & El-Mahallawi, I. S. (2019). Influence of process parameters in electrical discharge machining on H13 die steel. *Heliyon, 5*(6). https://doi.org/10.1016/j.heliyon.2019.e01813, e01813.

Bayramoglu, M., & Duffill, A. W. (1995). Manufacturing linear and circular contours using CNC EDM and frame type tools. *International Journal of Machine Tools and Manufacture, 35* (8), 1125–1136. https://doi.org/10.1016/0890-6955(95)90407-D.

D'Urso, G., Maccarini, G., Quarto, M., & Ravasio, C. (2015). Investigation on power discharge in micro-EDM stainless steel drilling using different electrodes. *Journal of*

Mechanical Science and Technology, 29(10), 4341–4349. https://doi.org/10.1007/s12206-015-0932-1.

D'Urso, G., Maccarini, G., Quarto, M., Ravasio, C., & Caldara, M. (2016). Micro-electro discharge machining drilling of stainless steel with copper electrode: The influence of process parameters and electrode size. *Advances in Mechanical Engineering, 8*(12), 1–16. https://doi.org/10.1177/1687814016676425.

Fukui, M., & Kinoshita, N. (1989). Developing a "mole" electric discharge digging machining. *CIRP Annals - Manufacturing Technology, 38*(1), 203–206. https://doi.org/10.1016/S0007-8506(07)62685-7.

Guirguis, & Kamal, S. (1993). *Electrical discharge machining of curved passages.* Rockwell International Corporation. NASA Tech Briefs.

Ids, M., & Clearfield, U. http://www.idsmachining.com/default.htm.

Ishida, T., Kogure, S., Miyake, Y., & Takeuchi, Y. (2004). Creation of long curved hole by means of electrical discharge machining using an in-pipe movable mechanism. *Journal of Materials Processing Technology, 149*(1–3), 157–164. https://doi.org/10.1016/j.jmatprotec.2003.11.043.

Ishida, T., Kogure, S., Miyake, Y., & Takeuchi, Y. (2005). Extension of curved hole length by means of electrical discharge machining using an in-pipe movable device. *Journal of the Japan Society for Precision Engineering, Contributed Papers, 71*(7), 911–915. https://doi.org/10.2493/jspe.71.911.

Ishida, T., Mochizuki, Y., & Takeuchi, Y. (2009). Elementary study on the creation of cross section's changing holes by means of electrical discharge machining. *International Journal of Automation Technology,* 592–601. https://doi.org/10.20965/ijat.2009.p0592.

Ishida, T., Nagasawa, H., Kita, M., & Takeuchi, Y.. (n.d.). Creation of cross-section changing hole with a hemisphere by means of electrical discharge machining (pp. 365–368). Springer Science and Business Media LLC. doi:https://doi.org/10.1007/978-1-84800-267-8_74.

Ishida, T., Nakajima, Y., Miyake, Y., & Takeuchi, Y. (2005). Development of electrode motion control device for electrical discharge curved hole machining. *Journal of the Japan Society for Precision Engineering, Contributed Papers, 71*(2), 262–266. https://doi.org/10.2493/jspe.71.262.

Ishida, T., Okahara, Y., Kita, M., Mizobuchi, A., Nakamoto, K., & Takeuchi, Y. (2014). Fundamental study on hole fabrication inside a hole by means of electrical discharge machining. *International Journal of Automation Technology, 8*(5), 773–782. https://doi.org/10.20965/ijat.2014.p0773.

Ishida, T., & Takeuchi, Y. (1999). Curved hole machining by self-movable mechanism with electrical discharge machining function. Development of automatic discharge gap controller. *Journal of the Japan Society for Precision Engineering, 65*(2), 245–249. https://doi.org/10.2493/jjspe.65.245.

Ishida, T., & Takeuchi, Y. (2002a). Development of electrode feed mechanism for electrical discharge curved hole machining using slider crank chains. *Journal of the Japan Society for Precision Engineering, 68*(2), 206–210. https://doi.org/10.2493/jjspe.68.206.

Ishida, T., & Takeuchi, Y. (2002b). L-shaped curved hole creation by means of electrical discharge machining and an electrode curved motion generator. *International Journal of Advanced Manufacturing Technology, 19*(4), 260–265. https://doi.org/10.1007/s001700200032.

Ishida, T., & Takeuchi, Y. (2008). Creation of U-shaped and skewed holes by means of electrical discharge machining using an improved electrode curved motion generator. *International Journal of Automation Technology,* 439–446. https://doi.org/10.20965/ijat.2008.p0439.

Ishida, T., & Takeuchi, Y. (2010). Design and implementation of automatic discharge gap controller for a curved hole creating microrobot with an electrical discharge machining function. *International Journal of Automation Technology*, 542–551. https://doi.org/10.20965/ijat.2010.p0542.

Ishiguro, E., Ishida, T., Kita, M., Nakamoto, K., & Takeuchi, Y. (2009). A11 development of CAD/CAM system for cross Section's changing hole electrical discharge machining : Formulation of post processor (digital design and digital manufacturing (CAD/CAM)). In *Proceedings of international conference on leading edge manufacturing in 21st century : LEM21* (pp. 251–256). https://doi.org/10.1299/jsmelem.2009.5.251.

Karunakaran, S. (2010). *Production technology*. Tata McGraw Hill Education Private Limited.

Kita, M., Ishida, T., Teramoto, K., & Takeuchi, Y. (2010). *Size reduction and performance improvement of automatic discharge gap controller for curved hole electrical discharge machining* (pp. 143–148). Springer Science and Business Media LLC. https://doi.org/10.1007/978-1-84882-694-6_25.

Lin, C. L., Lin, J. L., & Ko, T. C. (2002). Optimisation of the EDM process based on the orthogonal array with fuzzy logic and grey relational analysis method. *International Journal of Advanced Manufacturing Technology*, *19*(4), 271–277. https://doi.org/10.1007/s001700200034.

Meshram, D. B., & Puri, Y. M. (2019). Effective parametric analysis of machining curvature channel using semicircular curved copper electrode and OHNS steel workpiece through a novel curved EDM process. *Engineering Research Express*, *1*(1). https://doi.org/10.1088/2631-8695/ab337c.

Meshram, D. B., & Puri, Y. M. (2020). Optimized curved electrical discharge machining-based curvature channel. *Journal of the Brazilian Society of Mechanical Sciences and Engineering*, *42*(2). https://doi.org/10.1007/s40430-019-2162-4.

Miyake, Y., Ishida, T., & Takeuchi, Y. (2005). Electrical discharge machining of constant curvature curved hole by means of electrode motion control device. *Journal of the Japan Society for Precision Engineering, Contributed Papers*, *71*(11), 1388–1392. https://doi.org/10.2493/jspe.71.1388.

Mot, M., & Purcar, C. (2009). Development of curved hole machining method. *Nonconventional Technologies Review*, *4*, 77–80.

Nair, S., Dutta, A., Narayanan, R., & Giridharan, A. (2019). Investigation on EDM machining of Ti6Al4V with negative polarity brass electrode. *Null*, *34*(16), 1824–1831. https://doi.org/10.1080/10426914.2019.1675891.

Nakajima, Y., Ishida, T., Kita, M., Teramoto, K., & Takeuchi. (2006). *Development of curved hole machining method-size reduction of hole diameter-.Mechatronics for Safety, Security and Dependability in a New Era* (pp. 154–162). USA, Japan, China and Europe: Elsevier.

Okada, A., Yamaguchi, A., & Ota, K. (2017). Improvement of curved hole EDM drilling performance using suspended ball electrode by workpiece vibration. *CIRP Annals - Manufacturing Technology*, *66*(1), 189–192. https://doi.org/10.1016/j.cirp.2017.04.125.

Paswan, K., Pramanik, A., & Chattopadhyaya, S. (2020). Machining performance of Inconel 718 using graphene nanofluid in EDM. *Null*, *35*(1), 33–42. https://doi.org/10.1080/10426914.2020.1711924.

Praveen, L., Geeta Krishna, P., Venugopal, L., & Prasad, N. E. C. (2018). Effects of pulse on and off time and electrode types on the material removal rate and tool wear rate of the Ti-6Al-4V alloy using EDM machining with reverse polarity. *IOP Conference Series: Materials Science and Engineering*, *330*(1). https://doi.org/10.1088/1757-899X/330/1/012083. Institute of Physics Publishing.

Pushyanth, V. R. S., & Bhaskar, A. (2018). Experimental investigation and improvement of surface finish analysis on HCHCR AISI-D7 using EDM. *Materials Today: Proceedings*, *5*(5), 12115–12123. Elsevier Ltd https://doi.org/10.1016/j.matpr.2018.02.189.

Ramuvel, S. K., & Paramasivam, S. (2020). Study on tool steel machining with ZNC EDM by RSM, GREY and NSGA. *Journal of Materials Research and Technology*, *9*(3), 3885–3896. https://doi.org/10.1016/j.jmrt.2020.02.015.

Slatineanu, L., Alves-Coelho, A., Coteata, M., Uliuliu, Beşliu, I., & Mazuru, S. (2011). Teaching students the basics of designing experimental research experimental. In *Proceedings of ICAD. In The Sixth International Conference on Axiomatic Design* (pp. 195–203).

Straka, Ľ., & Hašová, S. (2018). Optimization of material removal rate and tool wear rate of cu electrode in die-sinking EDM of tool steel. *International Journal of Advanced Manufacturing Technology, 97*(5–8), 2647–2654. https://doi.org/10.1007/s00170-018-2150-3.

Uchiyama, M., & Shibazaki, T. (2004). Development of an electromachining method for machining curved holes. *Journal of Materials Processing Technology, 149*(1–3), 453–459. https://doi.org/10.1016/j.jmatprotec.2004.02.027.

Uliuliuc, D., Antonio, D., Coelho, M., & Slatineanuc, G. (2010). Electrical discharge machining of the curvilinear axis holes. *Nonconventional Technologies Review, 2,* 44–47.

Wang, X., Yi, S., Easton, M., & Ding, S. (2020). Active gap capacitance electrical discharge machining of polycrystalline diamond. *Journal of Materials Processing Technology, 280,* 116598. https://doi.org/10.1016/j.jmatprotec.2020.116598.

Yadav, V. K., Kumar, P., & Dvivedi, A. (2019). Performance enhancement of rotary tool near-dry EDM of HSS by supplying oxygen gas in the dielectric medium. *Materials and Manufacturing Processes, 34*(16), 1832–1846. https://doi.org/10.1080/10426914.2019.1675889.

Yamaguchi, A., Okada, A., & Miyake, T. (2015). Development of curved hole drilling method by EDM with suspended ball electrode. *Journal of the Japan Society for Precision Engineering, 81*(5), 435–440. https://doi.org/10.2493/jjspe.81.435.

CHAPTER THREE

Experimental and numerical investigation of deformation behavior of dual phase steel at elevated temperatures using various constitutive models and ANN

Sandeep Pandre[a], Ayush Morchhale[a], Nitin Kotkunde[a], Swadesh Kumar Singh[b], Navneet Khanna[c], and Ambuj Saxena[d]

[a]Department of Mechanical Engineering, BITS Pilani, Hyderabad, India
[b]Department of Mechanical Engineering, GRIET, Hyderabad, India
[c]Advanced Manufacturing Laboratory, Institute of Infrastructure Technology Research and Management (IITRAM), Ahmedabad, India
[d]G L Bajaj Institute of Technology and Management, Greater Noida, India

3.1 Introduction

The automotive industry is one of the most significant users of advanced high-strength steel (AHSS). It possesses all the suitable properties required for the construction of vehicle structures. DP steel is one of the grades of AHSS, which has a perfect blend of mechanical properties like ductility and high strength to weight ratio compared to other conventional steels. Its energy absorption capacity before failure makes it suitable for applications where crash performances are required. It finds its applications in the making of the bumper, wheel web, hoods, fenders, and some parts of the chassis (Kuziak, Kawalla, & Waengler, 2008). The microstructure of DP steel consists of martensitic and ferritic phases. The presence of dual-phase (ferrite and martensite) microstructure is the reason behind its exceptional mechanical properties, which are formed during thermomechanical processing (Zhao & Jiang, 2018). The material's high strength makes it challenging to form at normal ambient conditions, which can be overcome by implementing the elevated temperature forming. The flow stress of the material is lowered by applying heat, which makes the forming easier

(Pandre, Morchhale, Kotkunde, & Singh, 2020). The author, in his early work on DP steel, has studied the variation of flow stress behavior at elevated temperatures, i.e., from RT to 400°C. They found that the yield and ultimate stresses are reduced by 13.85 and 13.45% at 400°C compared to RT (Pandre et al., 2019). Cao, Karlsson, and Ahlström (2015) also studied the effect of temperature and strain rate on the deformation behavior of DP steels. The yield, ultimate tensile strength, and ductility were found to be varied at different test temperatures (-60 to $100°C$) and strain rates (10^{-4} to $10^{2}\,s^{-1}$).

At higher temperatures and strain rates, the material behaves in a nonlinear fashion, making it necessary for the prior prediction of its deformation behavior. Generally, during the experimental forming process, the parameters such as temperature, deformation rate, and strain govern the material plastic flow. The plastic flow behavior of the material is often predicted using various constitutive relations. It is used as an input for the numerical simulation for different manufacturing processes. Also, the accuracy of the numerical simulations largely depends on how accurately these constitutive relations are capturing the deformation behavior of the material. Several researchers have worked on the formulation of empirical, phenomenological, and physical-based constitutive relations to accommodate these requirements. Some of the frequently used constitutive models for the prediction of flow stresses are Hollomon (1945), Swift (1952), Voce (1948), Johnson and Cook (1985), and Zerilli and Armstrong (1987). For the accurate prediction of flow stress behavior behavior, the constitutive model should have a reasonable number of material constants evaluated during the experimentation data covering a more comprehensive range of temperature and strain rates. In the past, some investigators made changes in the original models and developed modified constitutive models. However, due to the empirical characteristics, their usage was limited for a range of strain rates and temperatures. To overcome the limitations, the phenomenological models can be replaced by physical-based models, which accurately capture the deformation behavior based on some physical assumptions (Lin, Chen, & Zhong, 2008). These models are based purely on physical assumptions with more material constants, which make it challenging to implement them in numerical simulations. The combined models used nowadays provide a better prediction compared to stand-alone models. Lin & Chen, 2010) developed a combined JC-ZA model to describe the relationship between the flow stress, strain rate, and temperature for hot-compressed 42CrMo alloy steel.

Earlier, some researchers have used artificial neural network (ANN) model to predict the flow stress behavior due to its ability to predict the nonlinear material responses at a wide range of temperatures and strain rates. One of the main advantages of this approach is that there is no requirement for mathematical function and material constants. Most recently, the ANN models were used to compare the predictive capabilities of available empirical, phenomenological, and physical-based models. Monajati, Asefi, Parsapour, and Abbasi (2010) investigated the effect of different process parameters on the mechanical properties and formability of deep drawing quality (DDQ) steel sheets. They have used a feed-forward back-propagation (FFBP) neural network for the prediction and found that the model could predict relevant process parameters with reasonable accuracy. Gupta, Anirudh, and Singh (2013) worked on the prediction of the flow stress of ASS 316 alloy at elevated temperatures (323–623 K) using four different models (JC, MZA, Arrhenius, and ANN). The prediction capability of different models was assessed using the correlation coefficient parameter. ANN showed the highest prediction compared to other models. Phaniraj and Lahiri (2003) worked on the applicability of ANN to low–carbon steels (0.03%–0.34% C). An FFBP network with a hyperbolic tangent was used as an activation function. The network has predicted the flow stress behavior with an average error of 3.7% of the mean flow stress. Sheikh and Serajzadeh (2008) estimated the flow stress behavior of AA5083 using ANN concerning the dynamic strain aging (DSA) effect. They have constructed two separate neural networks: the serrated-flow region and the other for smooth, yielding conditions. The predicted results were found to be in good agreement with the experimental data in both regions. Quan, Wang, Li, Zhan, and Xia (2016) also evaluated the dynamic flow stress of 7050 aluminum alloy in the temperature range of 573–723 K under a hot compression test using an FFBP learning algorithm. They found that the results of the well–trained ANN model were more accurate compared to the improved Arrhenius-type model. Reddy, Lee, Park, and Lee (2008) worked on predicting the flow stresses of different titanium alloys over a wide range of temperature and strain rates where the phase transformations involve. The BPNN learning algorithm with a sigmoidal function was used. They found that the ANN model can accurately predict the experimental data over a wide range of temperature and strain rates associated with the interconnection of metallurgical phenomena.

The present study deals with the experimental determination of the flow behavior of the as-received DP steel at elevated temperatures and different strain rates. Different constitutive models and neural network algorithms are used to predict the flow behavior over the studied temperature and strain rate range. Finally, a comparison is made between the different models for predicting the flow stress using statistical parameters like R, MAE, and SD.

3.2 Experimental procedure

3.2.1 Material characterization

In the present work, as-received DP steel alloy sheet of 1 mm thick was used for the experimentation. The material is characterized by performing a chemical composition test on the as-received material as per ASTM E415-15 standard. The chemical composition of the as-received DP steel sheet along with different weight percentages of the alloying elements present in it is mentioned in Table 3.1.

The microstructure of the DP steel was characterized by preparing the samples as per ASTM E3–95 standards. Firstly, the samples of 10 mm × 10 mm dimensions were cut from the as-received DP steel sheet along the RD of the sheet. Then, the samples were grounded on SiC paper with grit sizes of 240, 400, 600, and 1200 μm in a sequence. After grounding, the samples were polished on a disk polishing machine using three different grades of alumina powder until a mirror finish is obtained, i.e., starting with Grade-III and ending with Grade-I. Finally, the polished samples are etched using a reagent named Nital to reveal the grain boundaries present in the microstructure of the DP steel alloy. The composition of Nital consists of 10% of HNO_3 and 90% of methanol. The polished surfaces of DP 590 steel samples are swabbed slowly for 15–20 s using Nital reagent to reveal the grain boundaries clearly by avoiding corrosion during the etching process. The etched surfaces were rinsed with water, and moisture over the etched surfaces was removed. The microstructural examination was performed using a scanning electron microscope (SEM). The microstructure of the DP steel alloy in the RD at a magnification of 3500 × is shown in Fig. 3.1. The different phases present in the DP steel alloy were characterized through X-ray diffraction (XRD) analysis. The peaks in Fig. 3.2 reveal that

Table 3.1 Chemical composition of the as-received DP steel.

Alloying elements	C	Si	Mn	P	S	Cr	Mo	Ni	Fe
Weight (%)	0.075	0.26	2.29	0.007	0.003	0.45	0.3	0.006	Balance

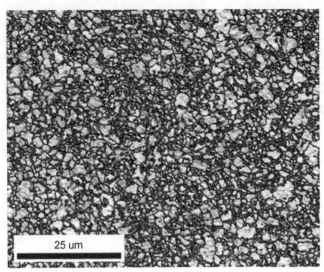

Fig. 3.1 The microstructure of the as-received DP steel sample.

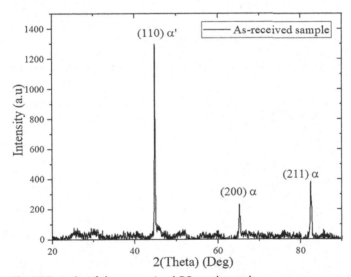

Fig. 3.2 The XRD peaks of the as-received DP steel sample.

the microstructure has two different phases, i.e., martensite and ferrite. The dark black part in the microstructure represents the martensitic phase. In contrast, the bright gray part shows the ferrite phase, and clear grain boundaries separate them.

3.2.2 Tensile testing

The specimens for tensile tests were prepared from the as-received sheet as per ASTM E8/E8M standard using CNC wire cut EDM as shown in Fig. 3.3. The uniaxial tensile tests were performed on a computer-controlled UTM as demonstrated in Fig. 3.4 at different temperatures (RT and 400 °C) and strain rates (0.0001, 0.001, and $0.01\,s^{-1}$). The machine has a loading capacity of 100 kN with a two-zone split furnace for high-temperature heating. The tests were repeated three times each, and the average values were considered to achieve repeatability.

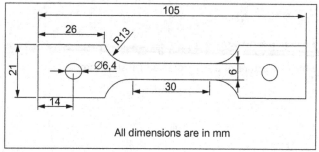

Fig. 3.3 The schematic of ASTM E8/E8M standard sub-sized tensile specimen.

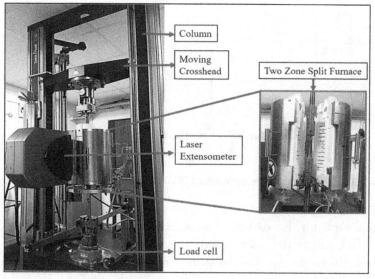

Fig. 3.4 The experimental setup for tensile testing.

3.3 Results and discussion

3.3.1 Experimental flow stress behavior

Figs. 3.5 and 3.6 show the representative plot of flow stress behavior of DP steel at different temperatures and fixed strain rates along the RD of the sheet. It can be observed from Figs. 3.5 and 3.6 that the temperature has a significant influence on the flow behavior of the material. Initially, the true stress of the material has shown a linear relation with the true strain for both the temperatures, after which the material has entered into the plastic deformation zone. In the plastic deformation regime, the material has experienced a high strain hardening in the common strain range of 0.04–0.15. The flow stresses have increased to a maximum level and reached a peak point known as the ultimate strength of the material. This region is also known as the uniform elongation region. After the maximum limiting point, neck formation has been observed due to which the material has undergone nonuniform elongation, followed by the failure of the specimen at some point. The flow stresses at 400°C are observed to be low compared to RT conditions at both strain rates. Due to the application of heat, thermal softening of the material takes place, which helps in the activation of slip systems at high temperatures, thereby allowing the free movement of dislocations with less restriction (Conrad, 1964).

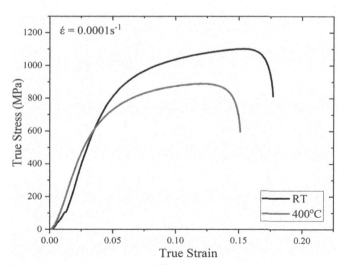

Fig. 3.5 True stress-strain graphs at different temperatures and at a strain rate of $0.0001 \, s^{-1}$.

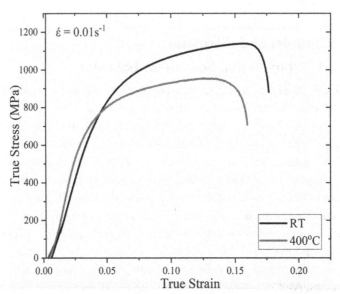

Fig. 3.6 True stress–strain graphs at different temperatures and at a strain rate of $0.01\,s^{-1}$.

Figs. 3.7 and 3.8 show the representative graphs at different strain rates and fixed temperatures. Figure 3.7 shows that with the change in strain rate from 0.0001 to $0.01\,s^{-1}$, the variation in flow stress has been observed. But, the variation in the flow stresses is not as significant as it can be seen at 400 °C shown in Fig. 3.8.

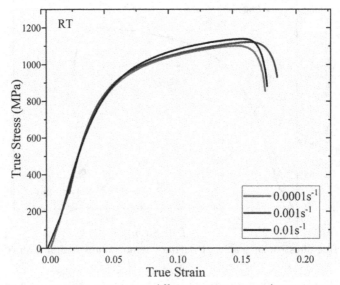

Fig. 3.7 True stress–strain response at different strain rates and room temperature.

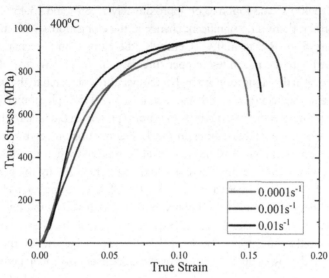

Fig. 3.8 True stress–strain response at different strain rates and 400°C temperature.

This can be due to the effect of temperature on the material's flow behavior at different strain rates. The effect of strain rate on the flow behavior is assessed by a parameter known as strain rate sensitivity (m), which is given in Eq. (3.1).

$$m = (\epsilon/\sigma)(d\sigma/d\epsilon) = ((\partial(ln\sigma))/(\partial(\ln \epsilon)))(\varepsilon, T = c) \qquad (3.1)$$

The higher value of "m" indicates that the material is more sensitive to strain rates at fixed strain and temperature conditions. The "m" values of the DP steel at RT and 400°C are observed to be 0.0070 and 0.0150, which indicates that the material's flow behavior is more sensitive at a higher temperature compared to RT. The various mechanical properties calculated based on the tensile tests are given in Table 3.2. The yield and ultimate tensile strength have shown an inverse relation with respect to temperature. The reduction in both the yield and ultimate strengths

Table 3.2 The mechanical properties of DP steel at different temperatures and strain rates.

Temperature (°C)	Strain rate (s^{-1})	Yield stress (MPa)	Ultimate stress (MPa)	Elongation (%)
RT	0.0001	715.00	942.14	19.10
	0.001	705.00	951.10	18.55
	0.01	725.00	970.00	17.91
400	0.0001	582.00	814.00	17.93
	0.001	650.00	863.28	18.37
	0.01	660.00	800.00	17.65

from RT to 400°C was found to be approximately 13.85% and 13.45%, respectively. Further, there is no significant change in the elongation% with the rise in temperature from RT to 400°C. Generally, the elongation percentage has to be increased with the rise in test temperature. But, in the present case, there is a slight decrease in the elongation at a higher temperature compared to the RT. This can be due to the occurrence of dynamic strain aging (DSA) phenomena, which causes the pinning of dislocations by solute atoms present in the metal. This slows down the movement of dislocations during the deformation process by generating a dragging force. The same phenomena will be repeated when the dislocations interact with the solute atoms. So, this could be the reason why the elongation has reduced at higher temperatures. And also, the DSA phenomenon is characterized by the negative strain-rate sensitivity, which is addressed by the author in his previous paper on flow stress behavior and microstructural characterization of DP steel at elevated temperatures (Pandre et al., 2019). Some literature in the past has also reported the loss of ductility caused by the DSA phenomenon in between the temperature range of 100–250°C for low-carbon steels (Taheri, MacCagno, & Jonas, 1995) and between 250 and 400°C for DP steels (Xiong, Kostryzhev, Stanford, & Pereloma, 2015).

The microstructure of the DP steel sample after the deformation at different test temperatures is shown in Figs. 3.9 and 3.10, respectively. It is evident from the microstructural analysis that the decrease in flow stress with respect to rise in temperature is due to the coarsening of grains present in it. The grain size has been found to be increased from 1.035 to 1.441 μm with the rise in test temperature from RT to 400°C, as shown in Figs. 3.9 and 3.10. Peaks in XRD analysis at different test conditions are shown in Figs. 3.11 and 3.12, which reveals that there is no change in the phases present in its microstructure even after testing at different temperatures. The presence of martensitic and ferrite phases for all the testing conditions can be inferred from XRD graphs except for the change in the intensity of the peaks. In DP steel alloy, the peaks corresponding to (110), (200), and (211) planes are observed. The peak at $2\theta = 44.803°$ is assigned to the martensite phase, and the peaks at $2\theta = 65.023°$ and 82.335° are assigned to ferrite phases. Similar observations were also reported by Xiong et al. (2015) in his work in DP steels.

Figs. 3.13 and 3.14 show the representative fractography of the DP steel samples tested at various temperatures and at a strain rate of $0.001\,S^{-1}$. The fractography has revealed the failure in DP steel samples at different temperature conditions to be ductile, which can be inferred from the presence of dimples. The dimples are formed due to the nucleation, growth, and coalescence of microvoids. In addition to these dimples, the presence of voids

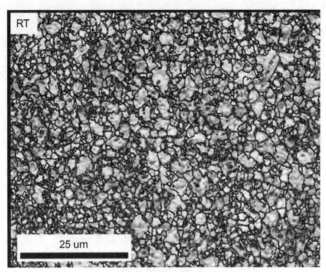

Fig. 3.9 Microstructure of the DP steel tested at room temperature.

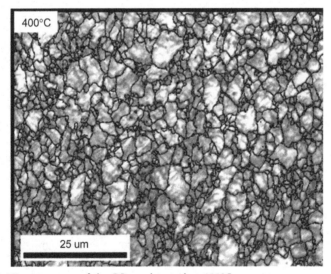

Fig. 3.10 Microstructure of the DP steel tested at 400°C.

larger than the size of the dimples was observed, especially for the samples tested at higher temperatures. These voids are initially smaller in size, and they become larger. This suggests that the formation of voids is due to the pullout of certain inclusions. In DP steels, the presence of such inclusions has been reported by Sirinakorn, Wongwises, and Uthaisangsuk (2014) in their work on the damage of dual-phase steel. The pullouts at the interface

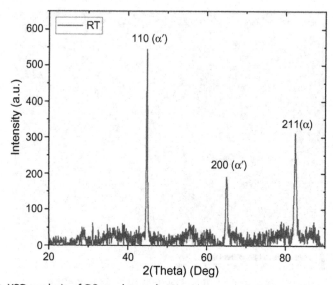

Fig. 3.11 XRD analysis of DP steel tested at room temperature.

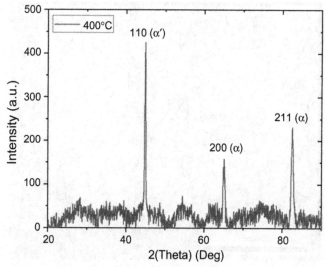

Fig. 3.12 XRD analysis of DP steel tested at 400°C.

between the martensite and ferrite act as a void nucleation site. Hence, in the present work, it can be concluded that the failure occurred due to a combined effect of formation of dimples, nucleation of microvoids, and inclusion pullouts. Interestingly, microvoid nucleation and inclusion pullout seem to be more dominant at higher temperature.

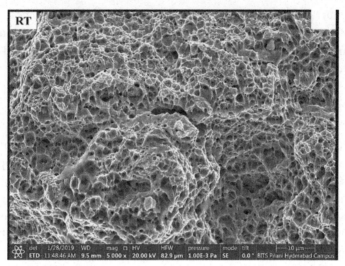

Fig. 3.13 Fractured surface of DP steel samples tested at room temperature.

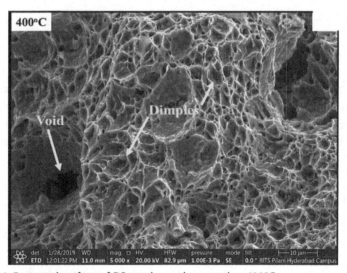

Fig. 3.14 Fractured surface of DP steel samples tested at 400°C.

3.3.2 Constitutive models

3.3.2.1 Modified Johnson-Cook (m-JC) model

The Johnson–Cook (JC) (Johnson & Cook, 1985) constitutive model is the most widely used constitutive model in predicting the flow stress behavior because of its simplicity with fewer number of material constants. It is a temperature, strain, and strain-rate-dependent phenomenological model and is successfully used for a variety of materials with different ranges of deformation temperature and strain rates. The original Johnson-Cook model can be expressed as shown in Eq. (3.2).

$$\sigma = \left(A + B\,\varepsilon^n\right)\left(1 + Cln\,\varepsilon^{\cdot}/\varepsilon^{\cdot}_{ref}\right)\left(1 - \left(\left(T - T_{ref}\right)/\left(T_m - T_{ref}\right)\right)^m\right)$$

(3.2)

The original JC model considers only the individual effects of strain hardening, strain-rate hardening, and thermal softening. It does not represent any thermal or strain-rate history effects. However, the coupling effects of the individual parameters are omitted in the original J-C model. Thus, to show the combined effect of the strain, strain rate, and temperature, some researchers like Zhang, Wen, and Cui (2009), Vural and Caro (2009), Shin and Kim (2010), and Lin, Chen, and Liu (2010) have made modifications to the original JC model. In the present study, the modified JC model developed by Lin et al. was considered as represented by Eq. (3.3).

$$\sigma = \left(A_1 + B_1\varepsilon + B_2\,\varepsilon^2\right)\left(1 + C_1\,ln\,\varepsilon^{\cdot}*\right)\exp\left[\left(\lambda_1 + \lambda_2\,ln\,\varepsilon^{\cdot}*\right)\left(T - T_r\right)\right]$$

(3.3)

where T_r is the reference temperature, T is the current temperature, ε is the strain, and $\dot{\varepsilon}^*$ is the ratio of current to the reference strain rate. It describes the coupled effects of strain rate and temperature on the flow stress behavior of different alloys. The material constants A_1, B_1, B_2, C_1, λ_1, and λ_2 for m-JC model have been determined using the procedure, followed by (Lin et al., 2010), and are represented in Table 3.3.

3.3.2.2 Modified Zerilli-Armstrong (m-ZA) model

The ZA model (Zerilli & Armstrong, 1987) works on the basis of various physical assumptions such as structure of the material (FCC, BCC, or HCP) and dislocation characteristics of the structure. The ZA model considers the coupled effect of strain and temperature effects. The ZA model is currently used as a part of various commercially available finite element software. However, researchers earlier reported that it has low prediction capability at a temperature above 0.6 times the melting point of material and at a very low deformation rate ($\sim$10–5/s). Hence, some major modifications have been suggested by Samantaray, Mandal, and Bhaduri (2009) and renamed it to modified Zerilli-Armstrong (m-ZA). The detailed method mentioned by Yuan, Li, Ji, Qiao, and Li (2014) in their work has been

Table 3.3 The calculated material constants for m-JC model.

A_1 (MPa)	B_1 (MPa)	B_2 (MPa)	C_1	λ_1	λ_2
702.68	2476.43	−748.65	0.0031	−0.0008	0.00045

followed step by step for calculating all the material constants of the m-ZA model and ultimately the flow stress using Eq. (3.4).

$$\sigma = (A + B\,\varepsilon^n)\exp\left\{(-C_3 + C_4\,\varepsilon)(T - T_{ref})\right. \\ \left. + \left[C_5 + C_6\left(T - T_{ref}\right)\ln\left(\dot{\varepsilon}/\dot{\varepsilon}_{ref}\right)\right.\right. \tag{3.4}$$

A, B, C_3, C_4, C_5, and C_6 are the material constants, and ε is considered as the strain at which the flow stress needs to be calculated. These constants are shown in Table 3.4.

3.3.2.3 Johnson-Cook and Zerilli-Armstrong (JC-ZA) model

The prediction capability of the flow stress can be improved by combining the individual constitutive models. In the above sections, only the individual phenomenological (m-JC) and physical-based (m-ZA) models are discussed. Here, in this section, a brief discussion is made on the combined JC-ZA. Nadai and Manjoine (1941) applied integrated JC-ZA model to different steels at elevated temperature and low deformation rate conditions. This model considers the effect of strain hardening and the yielding portion of the JC model and deformation rate and temperature terms from the ZA model. The main objective of this model is to have a coupled effect of deformation rate and temperature over the flow stress of metal. Thus, by considering the effect of plastic deformation, JC-ZA model is expressed using Eq. (3.5).

$$\sigma = (A + B\varepsilon^n)[-C_3(T + \Delta T) + C_4(T + \Delta T)\ln \dot{\varepsilon}^*] \tag{3.5}$$

The material constants A, B, C_3, C_4, and n are the material constants evaluated using the procedure, followed by Lin et al. (2008), and are given in Table 3.5.

Here, ΔT represents the average rise in the temperature during the plastic deformation of the metal due to the release of 3–5% deformation energy

Table 3.4 The calculated material constants for m-ZA model.

A (GPa)	B (GPa)	C_3	C_4	C_5	C_6	n
0.724	0.668	0.0057	0.0009	0.0072	0.0007	0.32

Table 3.5 The calculated material constants for JC-ZA model.

A	B	n	C_3	C_4
706.46	2548.56	0.31	0.0024	0.00014

stored inside the metal in terms of heat energy. In the present work, the average ΔT at RT (28°C) and 400°C deformation temperature has been recorded as 4 K and 15 K, respectively.

The JC-ZA model has also been considered in the present study because it considers the effect of yielding and strain hardening effect as well as the coupled effect of deformation rate and temperature while predicting the flow stress. Also, the JC-ZA model has an added advantage of fewer material constants compared to the m-JC model.

The ability of the different studied models m-JC, m-ZA, JC-ZA, and ANN models has been evaluated by comparing various statistical parameters (viz., correlation coefficient (R), mean absolute error (MAE), standard deviation (SD)). The selection of more than one statistical parameter helps to avoid biasing and improve the comparing efficiency. Figs. 3.15–3.17 show the representative graphs with a comparison of different statistical parameters for all the studied constitutive models. It can be seen from Fig. 3.15 that the m-JC model has shown a poor correlation coefficient ($R = 0.9324$) with the highest MAE and SD of 8.68% and 5.96% compared to other models. The other studied models like m-ZA and JC-ZA have shown a better correlation (R) value greater than 0.96 with MAE and SD less than 6% and 4% in comparison with the m-JC model shown in Figs. 3.16 and 3.17.

3.3.3 Artificial neural network (ANN)

A feed-forward back-propagation (FFBP) neural network consists of an input layer, an output layer, and one or more hidden layers depending on the dataset. In the FFBP network, the input data are segregated into training, testing, and validation data. The input is fed to the hidden layer and through the weighting function. The output of each hidden layer acts as an input to the next layer. The error in each layer is calculated and propagated back to the previous layers for corrections where the weights are adjusted accordingly. In the FFBP network, each processing unit in the output layer produces a single real number as its output, which is compared with the targeted output specified in the training set. This process is repeated until the weights of the first hidden layer are all adjusted. The flowchart of the whole FFBP process is shown in Fig. 3.18.

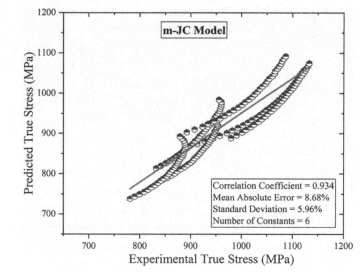

Fig. 3.15 The prediction capability of the m-JC constitutive model.

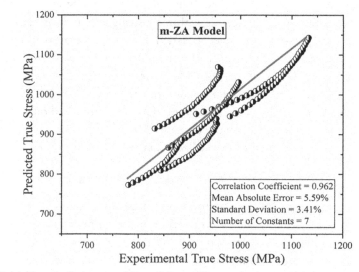

Fig. 3.16 The prediction capability of the m-ZA constitutive model.

A typical ANN in flow stress prediction consists of three layers, namely an input layer (strain, strain rate, and temperature), output layer (flow stress), and intermediate hidden layer (Prasad, Panda, Kar, Murty, & Sharma, 2019). The input and output layers are linked to the hidden layer by means of neurons. All the neurons of the network are connected by the weighted sum of

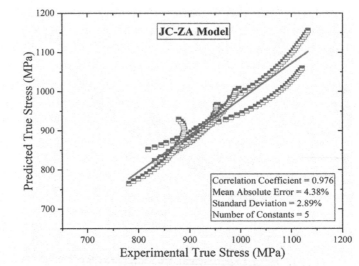

Fig. 3.17 The prediction capability of the JC-ZA constitutive model.

their input and output values (Morchhale, Kotkunde, & Singh, 2021). An ANN with a back-propagation algorithm is usually adopted for better prediction of the flow stress (Prasad et al., 2019). Hence, in the present work, an ANN structure with back-propagation neural networks was employed to model the flow behavior of DP. The schematic representation of the adopted ANN algorithm in the present work is shown in Fig. 3.19.

In order to make the efficient training network, input and output layer parameters were normalized in the domain of 0.05–0.95. The experimental data (x) were transformed to normalized data (x_n) using Eq. (3.6).

$$x_n = 0.05 + 0.9((x - x_{min})/(x_m ax - x_{min}))$$ (3.6)

where x_{min} and x_{max} are the minimum and maximum values of x, respectively. Apart from these, the determination of the number of intermediate layers is of chief importance to model the flow behavior accurately. The small number of intermediate layers in architecture may not have an adequate capacity to learn the process. On the other hand, a large number of hidden layers cause the overfitting of data (Prasad et al., 2019). To determine the optimum number of hidden layers, the mean square errors (MSEs) for various numbers of hidden layers were examined. It was observed that the least MSE has been observed when 10 neurons have been selected. Furthermore, the Levenberg-Marquardt function (*trainlm*) network was used to train the model. The developed ANN architecture consisted of an input

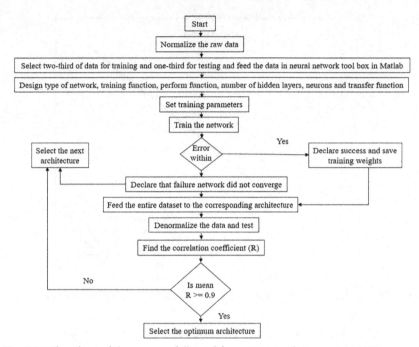

Fig. 3.18 Flowchart of the process followed for output prediction using ANN.

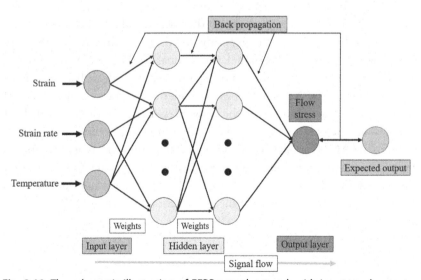

Fig. 3.19 The schematic illustration of FFBP neural network with inputs and outputs

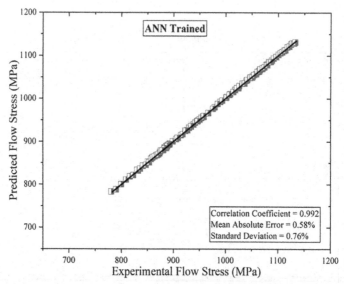

Fig. 3.20 The prediction capability of the ANN model for the trained dataset.

layer containing 3 neurons, a hidden layer containing 10 intermediate neurons, and an output layer containing 1 neuron.

Figs. 3.20 and 3.21 represent the prediction capability of the developed ANN model. The R values for the training and testing data were found to be 0.992 and 0.991, respectively. The mean absolute error and standard deviation values for the testing data were found to be around 1.13% and 1.38%, respectively. It was clearly evident that the trained and tested data fell almost on the same line as that of the experimental data, which indicated the very good suitability of the ANN model. On comparing the ANN predicted data with the considered constitutive models, it is clearly evident that the prediction capability of the ANN model was far superior to others. Thus, the ANN model can be used to describe the flow stress behavior of DP steel sheet material at the proposed testing temperature and strain-rate domain. Though the ANN model was best among all the four models, it was strongly dependent on an extremely good set of experimental data and offered no physical insight. Hence, the coupled JC–ZA model was considered to be the best suitable model to predict the deformation behavior of DP steel sheet material.

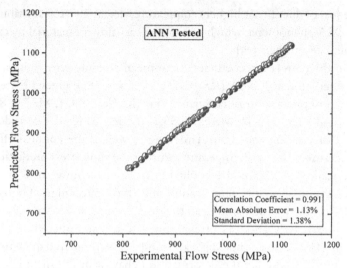

Fig. 3.21 The prediction capability of the ANN model for the tested dataset.

3.4 Conclusions

The current study investigates the experimental and prediction of flow stress behavior using four different models, viz. m-JC, m-ZA, JC-ZA, and ANN, for predicting the flow stress behavior of DP steel sheet at different strain rates and temperatures. The prediction capability of different models is compared using different statistical parameters. The following important conclusions have been drawn based on the abovementioned study:

- The flow stress of DP steel is more significantly influenced by temperature compared to strain rate. A decrease in the flow stress has been observed with the rise in temperature from RT to 400 °C, but no particular trend was observed with a change in the strain rates.
- The reduction in flow stress is due to the increase in the grain size from 1.035 to 1.441 μm with the rise in test temperature from RT to 400°C without change in the phases present in it. The presence of large-sized dimples in the fractography has also revealed the failure in the specimen to be ductile dominant.

- The loss of ductility at higher temperature, i.e., at 400°C, is attributed to the DSA phenomena, which restrict the easy flow of material by causing hindrances in their path.
- From the correlation coefficient, and mean absolute error values, it can be observed that JC-ZA ($R = 0.976$, MAE $= 4.38\%$) gives a better prediction of the flow stress compared to m-JC ($R = 0.934$, MAE $= 8.68\%$) and m-ZA ($R = 0.962$, MAE $= 5.59\%$) models, as JC-ZA considers the effect of yielding and strain hardening as well as the coupled effect of deformation rate and temperature during the flow stress prediction.
- The trained ANN model could track the exact flow stress behavior throughout the entire temperature and strain-rate range. Though the ANN model was best among all the three models, it was strongly reliant on an extremely good set of experimental data and offered no physical insight. Hence, the coupled JC-ZA model was considered to be the suitable model to predict the deformation behavior of DP sheet material.

References

Cao, Y., Karlsson, B., & Ahlström, J. (2015). Temperature and strain rate effects on the mechanical behavior of dual phase steel. *Materials Science and Engineering A, 636*, 124–132. https://doi.org/10.1016/j.msea.2015.03.019.

Conrad, H. (1964). Thermally activated deformation of metals. *JOM*, 582–588. https://doi.org/10.1007/BF03378292.

Gupta, A. K., Anirudh, V. K., & Singh, S. K. (2013). Constitutive models to predict flow stress in austenitic stainless steel 316 at elevated temperatures. *Materials and Design, 43*, 410–418. https://doi.org/10.1016/j.matdes.2012.07.008.

Hollomon, J. H. (1945). Tensile deformation. *Aime Trans, 12*, 1–22.

Johnson, G. R., & Cook, W. H. (1985). Fracture characteristics of three metals subjected to various strains, strain rates, temperatures and pressures. *Engineering Fracture Mechanics, 21*(1), 31–48. https://doi.org/10.1016/0013-7944(85)90052-9.

Kuziak, R., Kawalla, R., & Waengler, S. (2008). Advanced high strength steels for automotive industry. *Archives of Civil and Mechanical Engineering, 8*(2), 103–117. https://doi.org/10.1016/s1644-9665(12)60197-6.

Lin, Y. C., & Chen, X. M. (2010). A combined Johnson-Cook and Zerilli-Armstrong model for hot compressed typical high-strength alloy steel. *Computational Materials Science, 49*(3), 628–633. https://doi.org/10.1016/j.commatsci.2010.06.004.

Lin, Y. C., Chen, X. M., & Liu, G. (2010). A modified Johnson-Cook model for tensile behaviors of typical high-strength alloy steel. *Materials Science and Engineering A, 527*(26), 6980–6986. https://doi.org/10.1016/j.msea.2010.07.061.

Lin, Y. C., Chen, M. S., & Zhong, J. (2008). Constitutive modeling for elevated temperature flow behavior of 42CrMo steel. *Computational Materials Science, 42*(3), 470–477. https://doi.org/10.1016/j.commatsci.2007.08.011.

Monajati, H., Asefi, D., Parsapour, A., & Abbasi, S. (2010). Analysis of the effects of processing parameters on mechanical properties and formability of cold rolled low carbon steel sheets using neural networks. *Computational Materials Science, 49*(4), 876–881. https://doi.org/10.1016/j.commatsci.2010.06.040.

Morchhale, A., Kotkunde, N., & Singh, S. K. (2021). Prediction of flow stress and forming limits for IN625 at elevated temperature using the theoretical and neural network approach. *Materials Performance and Characterization, 10*(1), 146–165. https://doi.org/ 10.1520/MPC20200153.

Nadai, A., & Manjoine, M. J. (1941). High-speed tension tests at elevated temperatures— Parts II and III. *Journal of Applied Mechanics, 8*(2), A77–A91. https://doi.org/10.1115/ 1.4009105.

Pandre, S., Kotkunde, N., Takalkar, P., Morchhale, A., Sujith, R., & Singh, S. K. (2019). Flow stress behavior, constitutive modeling, and microstructural characteristics of DP 590 steel at elevated temperatures. *Journal of Materials Engineering and Performance, 28* (12), 7565–7581. https://doi.org/10.1007/s11665-019-04497-y.

Pandre, S., Morchhale, A., Kotkunde, N., & Singh, S. K. (2020). Influence of processing temperature on formability of thin-rolled DP590 steel sheet. *Materials and Manufacturing Processes, 35*(8), 901–909. https://doi.org/10.1080/10426914.2020.1743854.

Phaniraj, M. P., & Lahiri, A. K. (2003). The applicability of neural network model to predict flow stress for carbon steels. *Journal of Materials Processing Technology, 141*(2), 219– 227. https://doi.org/10.1016/s0924-0136(02)01123-8.

Prasad, K. S., Panda, S. K., Kar, S. K., Murty, S. V. S. N., & Sharma, S. C. (2019). Prediction capability of constitutive models for Inconel 718 sheets deformed at various elevated temperatures and strain rates. *Materials Performance and Characterization, 8*(5). https:// doi.org/10.1520/MPC20190004.

Quan, G.z., Wang, T., Li, Y.l., Zhan, Z.y., & Xia, Y.f. (2016). Artificial neural network modeling to evaluate the dynamic flow stress of 7050 aluminum alloy. *Journal of Materials Engineering and Performance, 25*(2), 553–564. https://doi.org/10.1007/s11665-016-1884-z.

Reddy, N. S., Lee, Y. H., Park, C. H., & Lee, C. S. (2008). Prediction of flow stress in Ti-6Al-4V alloy with an equiaxed $\alpha + \beta$ microstructure by artificial neural networks. *Materials Science and Engineering A, 492*(1–2), 276–282. https://doi.org/10.1016/j. msea.2008.03.030.

Samantaray, D., Mandal, S., & Bhaduri, A. K. (2009). A comparative study on Johnson Cook, modified Zerilli-Armstrong and Arrhenius-type constitutive models to predict elevated temperature flow behaviour in modified 9Cr-1Mo steel. *Computational Materials Science, 47*(2), 568–576. https://doi.org/10.1016/j.commatsci.2009.09.025.

Sheikh, H., & Serajzadeh, S. (2008). Estimation of flow stress behavior of AA5083 using artificial neural networks with regard to dynamic strain ageing effect. *Journal of Materials Processing Technology, 196*(1–3), 115–119. https://doi.org/10.1016/j. jmatprotec.2007.05.027.

Shin, H., & Kim, J.-B. (2010). A phenomenological constitutive equation to describe various flow stress behaviors of materials in wide strain rate and temperature regimes. *Journal of Engineering Materials and Technology, 132*(2). https://doi.org/10.1115/1.4000225.

Sirinakorn, T., Wongwises, S., & Uthaisangsuk, V. (2014). A study of local deformation and damage of dual phase steel. *Materials and Design, 64,* 729–742. https://doi.org/10.1016/j. matdes.2014.08.009.

Swift, H. W. (1952). Plastic instability under plane stress. *Journal of the Mechanics and Physics of Solids, 1*(1), 1–18. https://doi.org/10.1016/0022-5096(52)90002-1.

Taheri, A. K., MacCagno, T. M., & Jonas, J. J. (1995). Dynamic strain aging and the wire drawing of low carbon steel rods. *ISIJ International, 35*(12), 1532–1540. https://doi.org/ 10.2355/isijinternational.35.1532.

Voce, E. (1948). The relationship between stress and strain for homogeneous deformation. *Journal of the Institute of Metals, 74,* 537–562.

Vural, M., & Caro, J. (2009). Experimental analysis and constitutive modeling for the newly developed 2139-T8 alloy. *Materials Science and Engineering A, 520*(1–2), 56–65. https:// doi.org/10.1016/j.msea.2009.05.026.

Xiong, Z. P., Kostryzhev, A. G., Stanford, N. E., & Pereloma, E. V. (2015). Microstructures and mechanical properties of dual phase steel produced by laboratory simulated strip casting. *Materials and Design*, *88*, 537–549. https://doi.org/10.1016/j.matdes.2015.09.031.

Yuan, Z., Li, F., Ji, G., Qiao, H., & Li, J. (2014). Flow stress prediction of SiCp/Al composites at varying strain rates and elevated temperatures. *Journal of Materials Engineering and Performance*, *23*(3), 1016–1027. https://doi.org/10.1007/s11665-013-0838-y.

Zerilli, F. J., & Armstrong, R. W. (1987). Dislocation-mechanics-based constitutive relations for material dynamics calculations. *Journal of Applied Physics*, *61*(5), 1816–1825. https://doi.org/10.1063/1.338024.

Zhang, H., Wen, W., & Cui, H. (2009). Behaviors of IC10 alloy over a wide range of strain rates and temperatures: Experiments and modeling. *Materials Science and Engineering A*, *504*(1–2), 99–103. https://doi.org/10.1016/j.msea.2008.10.056.

Zhao, J., & Jiang, Z. (2018). Thermomechanical processing of advanced high strength steels. *Progress in Materials Science*, *94*, 174–242. https://doi.org/10.1016/j.pmatsci.2018.01.006.

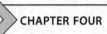

CHAPTER FOUR

Optimization of thermal efficiency of Scheffler solar concentrator receiver using slime mold algorithm

Anita Nene[a] and Omkar K. Kulkarni[b]
[a]Dr. Vishwanath Karad MIT World Peace University, Pune, India
[b]School of Mechanical Engineering, Dr. Vishwanath Karad MIT World Peace University, Pune, India

4.1 Introduction

Scheffler thermal concentrator be a milestone in solar thermal energy utilization. A Scheffler concentrator is a fixed focus concentrator, as reflected sunlight is focused on a fixed point where receiver is mounted and water is half-filled in the receiver. The generated steam is accumulated in the remaining half portion of receiver (Pavlović, Bellos, Stefanović, Tzivanidis, & Stamenković, 2016). To measure the performance of receiver in terms of thermal efficiency for various conditions such as shape of receiver, initial heating of inlet water up to 50°C in flat plate collector connected in series with Scheffler, tilting of receiver, glass cover to receiver surface. Cylindrical is the shape of receiver considered for experimentation (Panchal & Sadasivuni, 2020). Optimization study helps to study the parameter dependence on the performance of any system (Kumar, Prakash, & Kaviti, 2017). Optimization of thermal efficiency has been done by using response surface method and design expert software tool.

In recent decades, metaheuristic algorithms (MAs) have become popular in many applications because they have higher performance, lower computing power, and time required than deterministic algorithms in various optimization problems. Good results require a simple concept, and it is easy to transplant to different disciplines. In addition, some deterministic algorithms lack randomness in the later stage, and local optimal sinking often occurs. The random factor of MA enables the algorithm to find all optimal solutions in the search space, so local optimization can be effectively avoided. For linear problems, some gradient descent algorithms, for example, are more

Computational Intelligence in Manufacturing
https://doi.org/10.1016/B978-0-323-91854-1.00009-1
71

effective than stochastic algorithms in using gradient information. There are many algorithms that are nature-inspired such as cuckoo search algorithm (CSA) (Joshi, Kulkarni, Kakandikar, & Nandedkar, 2017), gray wolf algorithm (GWO) (Kulkarni & Kulkarni, 2018; Mirjalili, Mirjalili, & Lewis, 2014), whale optimization algorithm (WOA) (Seyedali Mirjalili & Lewis, 2016), Ant lion optimization algorithm (Mirjalili, 2015), grasshopper optimization algorithm (GOA) (Neve, Kakandikar, & Kulkarni, 2017), which prove their applications in the engineering field. Also, there are many socioinspired algorithms such as cohort intelligence (CI) (Kulkarni, Kulkarni, Kulkarni, & Kakandikar, 2018) and ideology algorithm (Huan, Kulkarni, Kanesan, Huang, & Abraham, 2017), which also benefit the use to engineering applications. This research deals with the application of a novel algorithm such as slime mold algorithm (SMA) (Li, Chen, Wang, Heidari, & Mirjalili, 2020) toward the thermal efficiency enhancement of Scheffler dish. Slime mold algorithm (SMA) has proven at a greater side with the results toward the thermal efficiency. Slime mold algorithm (SMA) is a unique nature-inspired algorithm.

4.2 Experimental setup

The system is fabricated as per the given specification. The main component of Scheffler collector system is Scheffler dish that consists of reflecting glass plates mounted on the typical elliptical frame and receiver placed at the focal point of Scheffler dish which supported by a stand (Azzouzi, Boumeddane, & Abene, 2017). Fig. 4.1 shows an experimental setup of Scheffler collector system, and the technical specification of the Scheffler collector system is given in Table 4.1.

The basic parameters recorded during the experiment are wind speed, total solar radiation, scattered solar radiation, humidity, and ambient temperature. K-type thermocouple is used to measure the surface temperature of the receiver (Ayub, Munir, Amjad, Ghafoor, & Nasir, 2018). Pressure and mass flow rate of steam generated were also measured to calculate thermal efficiency. Tilting arrangement was provided for the receiver, and measurement of tilting angle was done by using an inclinometer (Kumar et al., 2017). Experimentation was carried out at four different conditions as mentioned in Table 4.2.

4.3 Experimentation for thermal efficiency

Response surface methodology (RSM) is actually a set of statistical and mathematical methods used to build empirical models. By carefully

Fig. 4.1 Scheffler with cylindrical receiver. *No permission required.*

designing the experiment, the goal is to optimize the response (output variable) that is influenced by some independent variables (input variables). An experiment is a series of tests called execution, in which input variables are changed to determine why the output response has changed. The application of RSM in design optimization aims to reduce the cost and related numerical noise of expensive analysis methods (such as finite element methods and CFD analysis). The answer can be displayed graphically in three-dimensional space or as a contour map in order to visualize the shape of the answer area (Munir, Hensel, & Scheffler, 2010). Contours are constant response curves drawn in the x and y planes, and all other variables are fixed. Each contour line corresponds to a specific height of the response surface.

Table 4.1 Speciation of Scheffler.

Title	Scheffler reflector	Receiver	Tracking
Specification	Model—$2.7\,m^2$	(1) Cylindrical receiver	Manual tracking
	Major Axis—2200 mm	Diameter—450 mm	
	Minor Axis—830 mm	Length—50 mm	
	Concentration Ratio-16	(2) Volume—8 L	

Table 4.2 Design matrix for cylindrical receiver.

Conditions	Run	Factor 1 A: Average solar radiation (W/m²)	Factor 2 B: Mass flow rate of steam (kg/h)	Response 1 thermal efficiency (%)
General	1	417.00	0.90	52.19
condition	2	579.00	0.86	35.70
	3	665.00	0.63	22.45
Initial	4	691.00	1.60	56.64
heating	5	610.00	1.40	55.16
	6	526.00	1.19	53.61
Tilt effect	7	562.00	1.20	52.21
	8	552.00	0.80	34.83
	9	712.00	0.54	17.97
Glass	10	793.00	1.20	37.01
cover	11	865.00	0.84	23.34
	12	901.00	0.77	20.25

In this case, historical data design method is used for optimizing the thermal efficiency. Design-Expert 9.05 design tool software has been used for optimization.

4.3.1 Design of experiments

1. Create the design matrix in terms of input factor and output factor (response) by using experimental data set.
2. Analyze data for obtaining the correlation between input and output variables.

4.3.2 Analysis

1. Select the node and choose the transformation.
2. Fit the summary to evaluate the models for RSM.
3. Choose a significant effect from graph or list.
4. Select the model order and the desired conditions from the list.

4.3.3 Analysis of variance (ANOVA)

1. Analyze the selected model and see the results.
2. Evaluate model fit and transformation selection with charts.

In this case, optimization of thermal efficiency has been done for cylindrical receiver. For RSM, input factors are average solar radiation and mass flow rate of steam (Mendoza Castellanos, Carrillo Caballero, Melian Cobas, Silva

Lora, & Martinez Reyes, 2017). Thermal efficiency is the response for the given input variable of the receiver. From experimental readings, thermal efficiency is calculated, and the design of matrix is prepared as shown in Table 4.2. Experimental readings of all conditions are considered at 1 bar steam pressure.

For cylindrical receiver, correlation for input data and response are:
1. Thermal efficiency vs average solar radiation: -0.639
2. Thermal efficiency vs mass flow rate of steam: 0.843.
The negative correlation between solar radiation and thermal efficiency shows that thermal efficiency decreases with an increase in solar radiation (Chandrashekara & Yadav, 2017a, 2017b). The positive correlation between thermal efficiency and mass flow rate shows that thermal efficiency increases with an increase in mass flow rate. Analysis of model is performed by ANOVA as shown in Table 4.3.

4.3.4 General equation

$$\eta = (+ 57.91102) - (0.14917 \times I_{bn}) + (57.29927 \times M_s)$$
$$- (0.035226 \times I_{bn} \times M_s) + \left(9.43325 \times 10^{-5}\right) \times I_{bn}^{2}$$
$$+ 1.64911 \times M_s^{2.} \tag{1}$$

General equation is used to calculate the predicted value of the response. Graph shows a good agreement between actual value and predicted value of response with minimum residuals. RSM is used for finding out the equation for thermal efficiency (Aravindan, Giridharan, & Amirdesh Sudhan, 2018).

Table 4.3 ANOVA for response surface quadratic model for cylindrical receiver.

Source	Sum of squares	df	Mean square	F-value	P-value Prob > F	
Model	2472.77	5	494.55	989.25	< 0.0001	Significant
A	420.81	1	420.81	841.75	< 0.0001	
B	1406.76	1	1406.76	2813.92	< 0.0001	
AB	7.15	1	7.15	14.29	0.0092	
A^2	29.60	1	29.60	59.21	0.0003	
B^2	0.19	1	0.19	0.39	0.5560	
Residual	3.00	6	0.50			

The terms A and B are coded factors for average solar radiation and mass flow rate of steam, respectively. The model F-value of 989.25 implies that the model is significant. There is only a 0.01% possibility that this could end up in noise. Values of "Prob > F" less than 0.0500 indicate that the model terms are significant. In this case, A, B, AB, and A^2 are significant model terms. Values greater than 0.1000 indicate that the model terms are not significant (Daabo, Mahmoud, & Al-Dadah, 2016a, 2016b).

4.4 Slime mold optimization algorithm

This section will cover the fundamental principles and processes of slime mold. Then, based on its behavioral patterns, a mathematical model was created.

4.4.1 Originality

Some researchers have presented comparable naming algorithms prior to this article; however, the algorithm design and usage situations are substantially different from that methodology given in this article. Monismith and Mayfield addressed the single-objective optimization issue by modeling Dictyostelium amoeba's five life cycles: When using ANN to build a position-based initial grid, the nutritional state is considered, aggregation state, mound state, slug state, or decentralized state are all terms for the same thing. In Li and colleagues, the use of two types of slime tubular networks that conform to two different regional routing protocols is presented as a strategy for establishing wireless sensor networks (Li et al., 2020).

Combining the Physarum network with the ant colony system improves the algorithm's ability to avoid local optimum values, allowing it to better handle the traveling salesman issue. Schmickland Crailsheim presented a bionic navigation paradigm for swarm robots, inspired by the spread of slime bacteria. Becker created low-cost, fault-tolerant graphics by mimicking the slime mold *Physarum polycephalum*'s foraging behavior. The majority of modeling slime mold methods are employed in graph theory and generative networks, as can be observed from the previous description. The algorithm used to solve the optimization issue simulates Dictyostelium discoideum's five life cycles; however, the paper has fewer experiments and evidence.

The SMA suggested in this study mostly replicates the slime mold *P. polycephalum*'s behavior and morphological changes during foraging, but not its whole life cycle. At the same time, the usage of weights in SMA simulates the positive and negative feedback given by slime molds during the foraging process, resulting in the formation of three distinct morphologies, which is a novel concept. In addition, this study performs enough trials on the algorithm's properties. The SMA's superiority is demonstrated in the next section (Li et al., 2020).

4.4.2 Concept and elicitation

P. polycephalum is the slime mold that is referenced in this article. It was given the name "Slime mold" since it was initially classed as a fungus, and Howard described its life cycle in a 1931 study. Slime molds are eukaryotic creatures that prefer chilly, damp environments to thrive. The active and dynamic stage of Plasmodium and Myxomycetes, which is also the major study stage of this article, is the most important nutritional stage. At this point, the slime mold's organic substance searches for food, surrounds it, and generates enzymes to break it down.

The front end of the cell expands into a fan shape during migration, followed by a network of connecting veins that allow cytoplasm to flow within, as shown in Fig. 4.2. They may employ numerous food sources at the front end due to their distinctive modes and traits. A network of veins connects them all at the same time. The slime mold may reach a size of over 900 square centimeters if there is adequate food in the environment.

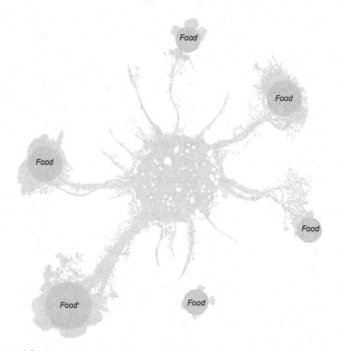

Fig. 4.2 Food-finding behavior of slime mold. *From Li, S., Chen, H., Wang, M., Heidari, A. A. & Mirjalili, S. (2020). Slime mould algorithm: A new method for stochastic optimization. Future Generation Computer Systems, 111, 300–323. https://doi.org/10.1016/j.future.2020. 03.055.*

Slime molds are commonly employed as model organisms due to their ease of cultivation on agar and oatmeal.

Kamiya and his colleagues were the first to investigate the cytoplasmic flow of slime molds in depth. Their research has greatly aided our knowledge of how slime mold spreads across the environment and links food. When the vein is close to the food supply, the bio-oscillator generates a spreading wave that enhances the flow of cytoplasm through the vein, we now know. The thicker the vein, the faster the cytoplasm flows. Slimes can construct the optimum path to connect food in a substantially superior fashion by combining positive and negative inputs. As a result, slime mold has been theoretically studied and used to route networks and graph theory.

Because the changes in vein structure and morphology evolve simultaneously with the contraction of slime mold, there are three connections between the deviations in vein structure and the contraction of slime mold.

(1) The thick veins grow approximately along the radius as the contraction frequency shifts from outside to inside.

(2) Anisotropy occurs when the contraction mode is unpredictable.

(3) The vein structure vanishes when the shrinking pattern of slime mold no longer organizes in time and space.

As a result, the link between slime mold vein structure and contraction mode matches the geometry of naturally generated cells. Flow feedback of the cytoplasm in the Physarum solver determines the thickness of each vein. The diameter of the vein expands as cytoplasmic flow increases. The vein shrinks as the flow lowers because the diameter lowers. Slime molds may create a more powerful pathway where the concentration of food is higher to ensure they absorb as many nutrients as possible (Li et al., 2020).

Slime mold has also been proven in recent investigations to be capable of arranging foraging according to optimal theory. Slime mold may pick the food source with the highest concentration when the quality of multiple food sources varies. Slime molds, on the contrary, must consider the speed and risk of foraging. Slime molds, for example, must make faster judgments to avoid being harmed by the environment. Experiments have revealed that the faster the slime mold makes decisions, the less probable it is to discover the major food supply. Slime molds must, therefore, consider speed and precision while selecting feeding sources. Slime molds must select whether to leave one location and look for food elsewhere when foraging.

Slime mold's best approach for estimating when to leave its present place when comprehensive information is absent is to apply heuristics or empirical principles based on the absence of currently accessible knowledge.

When slime mold comes into contact with high-quality food, its likelihood of leaving the region drops. Slime molds, on the contrary, may use many food sources at the same time due to their unique biological properties. As a result, even if the slime mold discovers a superior food source, it may still split a portion of the biomass to utilize both resources at the same time.

Slime mold may also change its search mode dynamically based on the quality of its food source. "When the quality of the food supply is high, the slime mold will employ the area restriction search strategy, narrowing the search to the food source that has been discovered. If the density of the first found food source is minimal, the slime mold will abandon it in search of other nearby food sources. This adaptive search approach can be better reflected when food bits of varying quality are dispersed in one location. In the next sections, several of the mechanisms and properties of the slime mold stated earlier will be mathematically represented" (Li et al., 2020).

4.4.3 Model based on mathematics

The suggested mathematical model and approach will be presented in depth in this section.

4.4.4 Approach toward food

Slime mold is attracted to food by odor in the air. The following formula is presented to mimic the contraction mode in order to represent its approximation behavior with a mathematical formula:

$$\overrightarrow{X(t+1)} = \begin{cases} \overrightarrow{X_b(t)} + \overrightarrow{vb} \cdot \left(\overrightarrow{W} \cdot \overrightarrow{X_A(t)} - \overrightarrow{X_B(t)} \right), r < p \\ \overrightarrow{vc} \cdot \overrightarrow{X(t)}, r \geq p \end{cases}$$

where $\overrightarrow{vb}$ is a parameter with a range of $[-a, a]$, $\overrightarrow{vc}$ decreases linearly from one to zero. t represents the current iteration, $\overrightarrow{X_b}$ represents the individual location with the highest odour concentration currently found, $\overrightarrow{X}$ represents the location of slime mould, $\overrightarrow{X_A}$ and $\overrightarrow{X_B}$ represent two individuals randomly selected from the swarm, $\overrightarrow{W}$ represents the weight of slime mould (Li et al., 2020)

The formula of p

$$p = \tanh |S(i) - DF|$$

Among them, $I = 1, 2, \ldots, n$, $S(i)$ denotes X's fitness and DF is the greatest fitness acquired over all iterations. $\overrightarrow{vb}$

$$\vec{vb} = [-a, a]$$

$$a = \text{arctanh}\left(-\left(\frac{t}{\text{max\_}t}\right) + 1\right)$$

The formula of $\vec{W}$

$$\overrightarrow{W(\text{Smell Index}(i))} = \begin{cases} 1 + r \cdot \log\left(\dfrac{bF - S(i)}{bF - wF} + 1\right), & \text{condition} \\[2mm] 1 - r \cdot \log\left(\dfrac{bF - S(i)}{bF - wF} + 1\right), & \text{others} \end{cases} \tag{2}$$

$$Smell\ Index = sort(S)$$

where condition denotes the top half of the population, $S(i)$ denotes a random value between 0 and 1, bF is the current iteration optimal fitness cc acquired in the current process, and Smell Index is a sequence sorting fitness value (rising in the minimum problem).

Fig. 4.3 depicts the effect of Eq. (2). The location of the sought individual X may be updated based on the best position presently acquired $\overrightarrow{xb}$ and the parameters vb, vc, and W may be fine-tuned to adjust the position of the person. Fig. 4.3 is also used to depict the search individual's location shift in three-dimensional space. The rand in the formula can cause the

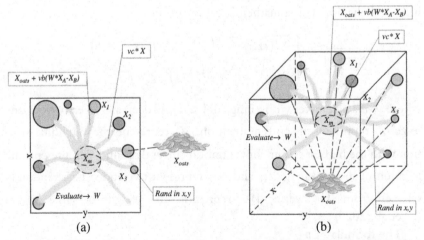

Fig. 4.3 Possible locations in 2 and 3 dimensions. *From Li, S., Chen, H., Wang, M., Heidari, A. A. & Mirjalili, S. (2020). Slime mould algorithm: A new method for stochastic optimization. Future Generation Computer Systems, 111, 300–323. https://doi.org/10.1016/j.future.2020. 03.055 possible locations in 2 and 3 dimensions.*

person to generate a search vector at any angle, that is, in any direction in the search solution space, so that the algorithm may locate the ideal solution. As a result of Eq. (1), the search individual will explore in all available directions toward the ideal solution, replicating the fan-shaped structure of slime mold while approaching food. This notion can also be extended to extra-dimensional space.

4.4.5 Wrap food

"The contraction of the slime mold's tissue structure during the search is mathematically simulated in this portion. The stronger the wave created by the bio-oscillator, the quicker the cytoplasm moves, and the thicker the vein, the higher the concentration of food in the vein. The positive and negative feedback between the vein width of slime mold and the food content under investigation are theoretically simulated by Eq. (5)" (Li et al., 2020). The components in Eq. (5) simulate the venous contraction pattern's uncertainty. The log is used to slow down the rate of change of the value, ensuring that the contraction frequency value does not fluctuate too much. Adjust their search mode according to the quality of the meal by simulating slime mold. "When the food concentration is high, the weight near the region increases; when the food concentration is low, the weight near the region decreases, allowing you to move on to other locations to explore. Fig. 4.4 depicts the procedure of determining the slime mold's fitness rating. The mathematical formula for updating the position of slime mold is as follows, based on the aforementioned principles" (Li et al., 2020):

$$\overrightarrow{X^*} = \begin{cases} \text{rand} \cdot (UB - LB) + LB, \ \text{rand} < z \\ \overrightarrow{X_b(t)} + \overrightarrow{vb} \cdot \left(W \cdot \overrightarrow{X_A(t)} - \overrightarrow{X_B(t)} \right), \ r < p \\ \overrightarrow{vc} \cdot \overrightarrow{X(t)}, \ r \geq p \end{cases}$$

The upper and lower bounds of the search range are represented by LB and UB, respectively, while random values in [0,1] are represented by rand and r. In the parameter setting experiment, the value of z will be addressed.

4.4.6 Grab food

The slime mold primarily relies on the biological vibrator's propagation wave to shift the cytoplasm flow in the vein, allowing them to be in a place with higher food concentration. W, $\overrightarrow{vb}$, $\overrightarrow{vc}$ are used to represent variations in the width of slime mold veins. W mathematically simulates the oscillation

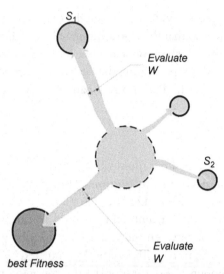

Fig. 4.4 Assessment of fitness. *From Li, S., Chen, H., Wang, M., Heidari, A. A. & Mirjalili, S. (2020). Slime mould algorithm: A new method for stochastic optimization. Future Generation Computer Systems, 111, 300–323. https://doi.org/10.1016/j.future.2020.03.055 Assessment of fitness.*

frequency of slime molds in the vicinity under different food concentrations, so slime molds can approach the food faster when finding high-quality foods and approach the food more slowly when the food concentration is low in a single location, increasing the viscosity. Pseudocode of SMA is presented in Fig. 4.5

Algorithm 1 Pseudo-code of SMA

Initialize the parameters *popsize, Max_iteraition*;
Initialize the positions of slime mould $X_i (i = 1,2, ..., n)$;
While (*t* ≤ *Max_iteraition*)
    Calculate the fitness of all slime mould;
    Update bestFitness, X_b
    Calculate the *W* by **Eq.** (5);
    For *each search portion*
      Update p, vb, vc;
      Update positions by **Eq.** (7);
    End For
    t = *t* + 1;
End While
Return *bestFitness, X_b*;

Fig. 4.5 Pseudocode of SMA. *No permission required.*

Table 4.4 Criteria for optimization for cylindrical receiver.

Name	Goal	Lower limit	Upper limit
Average solar radiation (Ibm)	is in range	417	901
Mass flow rate of steam (Ms)	is in range	0.54	1.60

The ranges of parameters of lower and upper bounds for the optimization problem are mentioned in Table 4.4.

Depending upon the criteria of optimization, 100 optimal solutions are generated, out of which 1 solution is selected. In this case, optimum efficiency is 60.05% with 95% confidence level.

4.5 Results and discussions

The experiments conducted with the 2.7-m^2 Scheffler system have identified the performance characteristics such as 52% of peak thermal efficiency, the best operating range up to 1 bar steam pressure, the best working temperature at 110°C, a steam generating rate of 0.9 kg/h, and a heat gain capacity of 0.556 kW. The small Scheffler concentrator has a limited reflector area and small concentration ratio; therefore, it cannot perform well beyond the 1–1.5 bar pressure range in its normal operation state. For initial heating conditions, the efficiency is 56% and the mass of steam generation rate is 1.6 kg/h experimentally.

The parametric analysis identified the initial feed water heating as the most effective measure for cylindrical receiver efficiency improvement. The response surface methodology (RSM) technique was used to determine the design factor setting for optimizing the thermal efficiency of the Scheffler dish. The design factors were steam mass flow rate and solar radiation.

The cylindrical receiver showed an optimum thermal efficiency value of 60.05% with initial heating at 1 bar steam pressure, solar radiation of 611.5 W/m^2, and a mass flow rate of 1.52 kg/h by RSM technique as shown in Fig. 4.6.

On solving Eq. (1) for maximization of thermal efficiency with consideration of upper and lower bounds as mentioned in Table 4.4, we get the best optimum value for the response variable as efficiency and best parameter settings. The results out of the algorithm were quite promising with respect to the efficiency output. The performance of the slime mold algorithm has already proved its superiority in the application to real life. Table 4.5 indicates the optimized values from the algorithm and parameters as well.

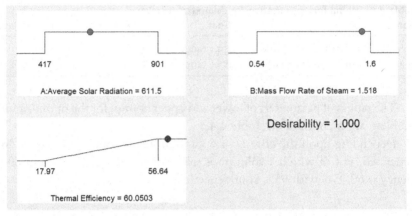

Fig. 4.6 Schematic representation of ramps. *From Li, S., Chen, H., Wang, M., Heidari, A. A. & Mirjalili, S. (2020). Slime mould algorithm: A new method for stochastic optimization. Future Generation Computer Systems, 111, 300–323. https://doi.org/10.1016/j.future. 2020.03.055.*

Table 4.5 Optimized results from SMA.

Name	Optimized parameters	Thermal efficiency
Average solar radiation (Ibm)	417	84%
Mass flow rate of steam (Ms)	1.60	

4.6 Conclusion

Experimental analysis results show that the thermal efficiency of the cylindrical receiver is maximum at 1 bar steam pressure for initial heating. As solar radiation changes thermal efficiency also changes; if solar radiation increases above the specified limit as per the guidelines given by the Bureau of Indian Standards, loss from receiver increases and efficiency decreases. Experimentally, it is difficult to maintain constant solar radiation at $611.5\,\text{W/m}^2$ due to the change of environmental conditions. As per the RSM technique, the optimum condition is solar radiation of $611.5\,\text{W/m}^2$ and a mass flow rate of $1.52\,\text{kg/h}$, and optimum thermal efficiency is 60.05% and the condition must be initial heating of water. But with the application of the slime mold algorithm, we get an even better combination of the parameters with the initial condition as the same. The slime mold algorithm gives us the solar radiation and mass flow rate as $417\,\text{W/m}^2$ and $1.6\,\text{kg/h}$, respectively. The efficiency of the system can be increased by these parameter settings by almost 39%. This is a drastic change in the performance of the system.

References

Aravindan, S., Giridharan, N., & Amirdesh Sudhan, B. (2018). The recent trends in optimization of thermal performance of parabolic trough solar collector. *International Journal of Mechanical Engineering and Technology*, *9*(5), 655–661. https://www.iaeme.com/MasterAdmin/Journal_uploads/IJMET/VOLUME_9_ISSUE_5/IJMET_09_05_072.pdf.

Ayub, I., Munir, A., Amjad, W., Ghafoor, A., & Nasir, M. S. (2018). Energy- and exergy-based thermal analyses of a solar bakery unit. *Journal of Thermal Analysis and Calorimetry*, *133*(2), 1001–1013. https://doi.org/10.1007/s10973-018-7165-3.

Azzouzi, D., Boumeddane, B., & Abene, A. (2017). Experimental and analytical thermal analysis of cylindrical cavity receiver for solar dish. *Renewable Energy*, *106*, 111–121. https://doi.org/10.1016/j.renene.2016.12.102.

Chandrashekara, M., & Yadav, A. (2017a). An experimental study of the effect of exfoliated graphite solar coating with a sensible heat storage and Scheffler dish for desalination. *Applied Thermal Engineering*, *123*, 111–122. https://doi.org/10.1016/j.applthermaleng.2017.05.058.

Chandrashekara, M., & Yadav, A. (2017b). Experimental study of exfoliated graphite solar thermal coating on a receiver with a Scheffler dish and latent heat storage for desalination. *Solar Energy*, *151*, 129–145. https://doi.org/10.1016/j.solener.2017.05.027.

Daabo, A. M., Mahmoud, S., & Al-Dadah, R. K. (2016a). The effect of receiver geometry on the optical performance of a small-scale solar cavity receiver for parabolic dish applications. *Energy*, *114*, 513–525. https://doi.org/10.1016/j.energy.2016.08.025.

Daabo, A. M., Mahmoud, S., & Al-Dadah, R. K. (2016b). The optical efficiency of three different geometries of a small scale cavity receiver for concentrated solar applications. *Applied Energy*, *179*, 1081–1096. https://doi.org/10.1016/j.apenergy.2016.07.064.

Huan, T. T., Kulkarni, A. J., Kanesan, J., Huang, C. J., & Abraham, A. (2017). Ideology algorithm: A socio-inspired optimization methodology. *Neural Computing and Applications*, *28*, 845–876. https://doi.org/10.1007/s00521-016-2379-4.

Joshi, A. S., Kulkarni, O., Kakandikar, G. M., & Nandedkar, V. M. (2017). Cuckoo search optimization—A review. *Materials Today: Proceedings*, *4*(8), 7262–7269. Elsevier Ltd https://doi.org/10.1016/j.matpr.2017.07.055.

Kulkarni, O., & Kulkarni, S. (2018). Process parameter optimization in WEDM by grey wolf optimizer. *Materials Today: Proceedings*, *5*(2), 4402–4412. Elsevier Ltd https://doi.org/10.1016/j.matpr.2017.12.008.

Kulkarni, O., Kulkarni, N., Kulkarni, A. J., & Kakandikar, G. (2018). Constrained cohort intelligence using static and dynamic penalty function approach for mechanical components design. *International Journal of Parallel, Emergent and Distributed Systems*, *33*(6), 570–588. https://doi.org/10.1080/17445760.2016.1242728.

Kumar, A., Prakash, O., & Kaviti, A. K. (2017). A comprehensive review of Scheffler solar collector. *Renewable and Sustainable Energy Reviews*, *77*, 890–898. https://doi.org/10.1016/j.rser.2017.03.044.

Li, S., Chen, H., Wang, M., Heidari, A. A., & Mirjalili, S. (2020). Slime mould algorithm: A new method for stochastic optimization. *Future Generation Computer Systems*, *111*, 300–323. https://doi.org/10.1016/j.future.2020.03.055.

Mendoza Castellanos, L. S., Carrillo Caballero, G. E., Melian Cobas, V. R., Silva Lora, E. E., & Martinez Reyes, A. M. (2017). Mathematical modeling of the geometrical sizing and thermal performance of a dish/Stirling system for power generation. *Renewable Energy*, *107*, 23–35. https://doi.org/10.1016/j.renene.2017.01.020.

Mirjalili, S. (2015). The ant lion optimizer. *Advances in Engineering Software*, *83*, 80–98. https://doi.org/10.1016/j.advengsoft.2015.01.010.

Mirjalili, S., & Lewis, A. (2016). The whale optimization algorithm. *Advances in Engineering Software*, *51–67*. https://doi.org/10.1016/j.advengsoft.2016.01.008.

Mirjalili, S., Mirjalili, S. M., & Lewis, A. (2014). Grey wolf optimizer. *Advances in Engineering Software*, *69*, 46–61. https://doi.org/10.1016/j.advengsoft.2013.12.007.

Munir, A., Hensel, O., & Scheffler, W. (2010). Design principle and calculations of a Scheffler fixed focus concentrator for medium temperature applications. *Solar Energy*, *84*(8), 1490–1502. https://doi.org/10.1016/j.solener.2010.05.011.

Neve, A. G., Kakandikar, G. M., & Kulkarni, O. (2017). Application of grasshopper optimization algorithm for constrained and unconstrained test functions. *International Journal of Swarm Intelligence and Evolutionary Computation*. https://doi.org/10.4172/2090-4908.1000165.

Panchal, H., & Sadasivuni, K. K. (2020). Investigation and performance analysis of Scheffler reflector solar cooking system integrated with sensible and latent heat storage materials. *International Journal of Ambient Energy*, *41*(10), 1096–1105. https://doi.org/10.1080/01430750.2018.1501754.

Pavlović, S. R., Bellos, E., Stefanović, V. P., Tzivanidis, C., & Stamenković, Z. M. (2016). Design, simulation, and optimization of a solar dish collector with spiral-coil thermal absorber. *Thermal Science*, *20*(4), 1387–1397. https://doi.org/10.2298/TSCI160213104P.

CHAPTER FIVE

Study on drilling behavior of polymer nanocomposites modified by carbon nanomaterial with fiber: A case study

Jogendra Kumar, Kesarwani Shivi, Balram Jaiswal, Kaushlendra Kumar, Devendra Kumar Singh, Kuldeep Kumar, Rahul Vishwakarma, and Rajesh Kumar Verma

Materials and Morphology Laboratory, Department of Mechanical Engineering, Madan Mohan Malaviya University of Technology, Gorakhpur, Uttar Pradesh, India

5.1 Introduction

Materials are essential parts of our modern society to develop different types of components and structures. The mankind process of the material took place across many significant phases, such as Stone Age, Bronze Age, and Iron Age (the Industrial Revolution). Many synthetic polymeric materials were developed in industrial chemistry after the Second World War, and therefore now we live in Polymer Era (Layth et al., 2015; Wang et al., 2011). In the 21st century, hybrid polymeric materials play a primary function in the production sector due to improved properties like high strength to load proportion, anti-corrosive features, dimensional tolerances, high elasticity, damping coefficient, etc., (Cui et al., 2019). The development of novel lightweight composites with improved mechanical features leads to synthesizing a new generation of composites. Polymer composites are highly exploited in the components of aerospace, naval, space, and automotive productions with better mechanical properties and modified fatigue life (Brebu, 2020; Uygunoglu et al., 2015). In addition to thermoset polymers, cross-linked connection contributes to higher stiffness, thermal and mechanical properties. The cross-linked nature, anisotropic, and non-homogeneity of fiber composite create an enhancement of engineering properties. Sometimes, these features generate some machining challenges during manufacturing operations. It can be controlled through the optimal set of process parameters. The simultaneous control of the varying parameters can limit these challenges through the application of optimization modules.

5.1.1 Carbon nanomaterials

In the epoxy resin, reinforcing of nanofiller materials was remarked as an effective technique to boost the strength and stiffness under different loading conditions. The supplement of nanomaterials in the epoxy matrix leads to the augmentation of the flexural and tensile behavior. The previous study beheld the use of carbon nanomaterials (fullerenes, nanotubes (NT), nanorods (NR), and graphene) for the enhancement of composite mechanical properties (Burkov & Eremin, 2018; Gang, 2017). The broad classification of carbon nanomaterials is explained in Fig. 5.1.

5.1.2 Polymer nanocomposite

Composite laminates are becoming accessible in several technological fields, mainly in the automobile industry, sports equipment, aerospace products,

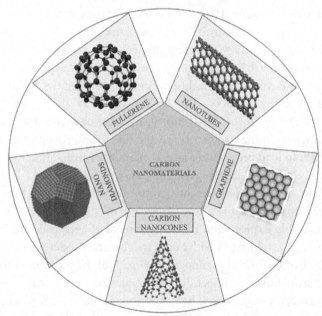

Fig. 5.1 Classification of carbon nanomaterials. *From Rauti, R., Musto, M., Bosi, S., Prato, M., & Ballerini, L. (2019). Properties and behavior of carbon nanomaterials when interfacing neuronal cells: How far have we come?* Carbon, 143, 430–446. https://doi.org/10.1016/j.carbon.2018.11.026.

ship marine, and petroleum trades, due to their novel mechanical and physical characteristics (Layth et al., 2015; Wang et al., 2011). The dominant approach in the study of fiber-modified polymers are glass-, carbon-, and kevlar-reinforced composites. It focuses on using products with different assets (Nijssen, 2015). The new development in aviation designs recommends a weight reduction to around 50% by substituting the main structural elements with manufactured FRP composites (Kumar et al., 2019; Myagkov et al., 2014). To accomplish reduced fuel consumption and improve engine performance objectives of future growth aircraft design technologies, the use of lightweight and elevated strength composites is essential (Anh Nguyen et al., 2019; Stewart, 2009). Polymer composites become very common for a more significant number of functions due to their excellent mechanical properties, comprising low weight, stiffness to weight, resistance to corrosion, fatigue, and thermal coefficient compared with metals (Anh Nguyen et al., 2019; Myagkov et al., 2014; Stewart, 2009).

Moreover, by carefully altering carbon fiber ratios, the fiber orientation of individual ply in laminates and material properties can be modified to suit various needs (Liu et al., 2012). Based on the selected materials and manufacturing processes, the reinforced element comprises particles, flakes, whiskers, laminates, or fibers (George et al., 2004; Layth et al., 2015). Carbon, glass, graphite, and aramid are the most widely used fibrous reinforcement (Papageorgiou et al., 2019). Due to its improved durability and process simplicity regarding other traditional products, polymer matrix composites have attracted considerable attention among material professionals. The macro-reinforced (fiber) composite existing properties are prominently improved through the supplement of nanomaterials. Doping theory is sometimes used in fiber composites to investigate polymer physical and mechanical aspects (Bal & Saha, 2014; Kumar & Kumar Verma, 2020; Saadatmandi et al., 2020). Limited studies have been done on the use of nanomaterials in carbon/epoxy composites. It requires more attention for the efficient utilization of polymer nanocomposites for multifunctional products.

Fig. 5.2 explains the major challenges of finding a suitable structural material for fuel economy (Sun & Zhao, 2020). The ASHBY chart shows the strength vs. density plot for material selection, which indicates that metals have the highest strength but are not a highly suitable structural material due to their higher density. On the other hand, carbon fiber-reinforced polymer (CFRP) has a lower density and higher strength. The induction of nanomaterials into the laminates composite has significantly enhanced the desired properties with a high synergistic effect (Ashby et al., 2004;

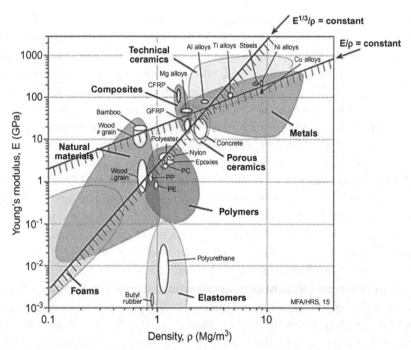

Fig. 5.2 Material selection chart (ASHBY chart). *From Shercliff, H. R., & Ashby, M. F. (2016). Elastic structures in design. Elsevier. https://doi.org/10.1016/B978-0-12-803581-8.02944-1.*

Findik & Turan, 2012). The nanomaterials such as carbon nanotube (CNT), carbon nano-onions (CNO), graphene nano-palates (GNP), reduced graphene oxide (rGO), graphene oxide (GO), etc., are broadly used in macrolevel polymer composite (Kato et al., 2012). Earlier investigation reveals that carbon-based nanomaterials (CNMs) such as GO, rGO, CNT, multiwall carbon nanotube (MWCNT), etc., show wide ranges of properties and higher application potentials in the polymer matrix (Cao et al., 2020; Changgu et al., 2008; Ray & Easteal, 2007). Thus, the practical application of graphene nanomaterials into CFRP increases its mechanical strength significantly with a lower density (Kwon et al., 2017; Pathak et al., 2016). The weight of components made with carbon fiber was decreased by 60% compared to steel and 30% to aluminum (Ashby et al., 2004; Ashby & Ashby, 2013).

5.2 Machining process (drilling) on laminated polymer nanocomposite

The machining of carbon-reinforced polymers is substantially different from metallic materials and their mixtures. The machining phenomenon of metals is not applicable for polymers and their composites. Generally, the fabricated polymer composite components are produced in a net near shape. However, still, there is a requirement of a secondary manufacturing process for fitting and joining into the final assembly of different products (Fleischer et al., 2018; López et al., 2011; Slamani et al., 2015). The complex structure of matrix composition and the high abrasiveness of the reinforcement material is challenging during machining polymer composites. Machining-induced defects and damage may obstruct cutting forces and tool life. The machining operation (drilling) of a composite is highly reliant on the properties of the reinforced material, matrix phase (as mentioned in Fig. 5.3). Investigations reveal that the proper control of machining constraints could effectively lower damages and defects generation during machining Voss et al., 2017; Ozkan et al., 2019; Komanduri, 1997). Due to its complex microstructure, manufacturer and practitioner-researchers face crucial challenges when machining polymers reinforced by carbon nanomaterials (CNMs) like carbon nanorods/nanotubes/graphene/reduced graphene oxide, etc. The tool life and balancing of cutting conditions for product manufacturing are complex tasks for the manufacturing sector and academia (Allwin Roy et al., 2017; Sengupta et al., 2011).

There is an extensive need to control the splintering, delamination, cracking, burr formation, fiber pullout, etc., (Liu et al., 2012; Voß et al., 2014).

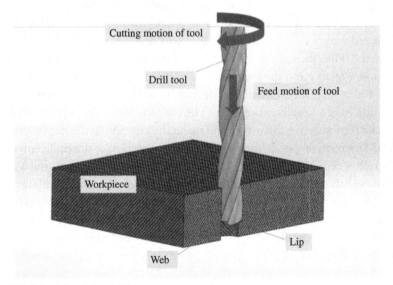

Fig. 5.3 Principle of the drilling process.

The different features of the fiber and matrix have a major effect on the processing by fiber orientation. It is mainly due to complex and abrasive fibers and soft matrix; the appropriate cutting conditions are difficult (Gaugel et al., 2016). Thrust and torque forces were monitored during the drilling of laminated composites (Anarghya et al., 2018). Based on observational findings, composite materials are subjected to lesser plastic distortion and a crack resistance 10 to 100 times lower than ordinary steels (Adak et al., 2019). In the past few years, customer demands emphasized developing amended machining technology with vital manufacturing dealing issues. The closeness form in a structure according to design specifications is one of the main advantages of composites. As a result, an innovative methodology for polymer machining is necessary to achieve high productivity and lower costs (Liu et al., 2012). Machining is essential for various stages of manufacturing, such as parts production and assembly. Underprivileged machined quality leads to poor assembly and a fast-term degradation of structural efficiency (Ho-Cheng & Dharan, 1990; Khashaba, 2004). Hole processing is one of the most frequent methods for secondary composite machining because of the need for mechanical parts and assemblies to rivet and bolt (Bielawski et al., 2017). Several operations are investigated, such as CNC drilling, laser beam drilling, ultrasound drilling (Dahnel et al., 2016; Santhanakrishnan Balakrishnan et al., 2019). However, the most economical and effective procedure of the composite tends to be traditional drilling using a twist drill. Still, a limited attention is paid to the drilling technique, while up to 40% of the process is devoted to hole making, as revealed by a medium-sized industrial report (Giasin et al., 2016). However, compared with a straight-edge drill, the twist drill has a moderately complex geometry. The efficiency is different, with the outer diameter of the drill being the most effective and the least efficient in its central area (Hocheng & Tsao, 2003; Liu et al., 2012). The edge of the chisel and the lips near the center of the twist drill have a negative rake angle. The outcome of a large negative angle of the rake is to improve this action and make the creation of the chip more difficult. The relative speed drops linearly in the direction of the drill center and approaches zero. This, in turn, restricts its performance in generating a hole. Consequently, the materials that penetrate the chisel edge of the drilling point into the hole are expected to be extruded rather than cut. There is a high force of thrust to move the twist drill through the work and the heat produced damage to the chisel edge of the drill tool (El-Sonbaty et al., 2004; Saoudi et al., 2016).

5.2.1 Machining-induced damages in laminated polymer nanocomposite

Composite damage, termed delamination, represents a drilling-induced image processing condition that is a predominant problem and has been described in the drilling of composite laminates as severe and unexpected damage (John et al., 2018). Because of the non-homogeneity of composites, the multiple phase structure and anisotropic nature lead to interruption of the drilling process. Delamination may occur due to uncut composite laminates in the inlet and outlet section. Another damage mechanism is subsurface deformation caused by the drilling test of polymers (Gara et al., 2018; Haeger, Schoen, et al., 2016). Matrix deformation and splitting, interfacial debonding, fiber pullouts, and spalling are examples of substructure defects (Haeger, Meinhard, et al., 2016). Thus, the material defects of delaminating holes should be corrected by an appropriate range of cutting constraints, drill geometries, drill sizes, and cutting environments to improve product efficiency and durability (Bhatnagar et al., 2004; Hocheng & Tsao, 2003). However, structural strength mainly depends on interfacial connections with the fiber matrix and fiber patterns (Kumar & Singh, 2017). Fig. 5.4 displays the development of defects, and approximately 60% of the total part rejection is of poor machined quality, and part rejection is costly, resulting in low product quality (Aamir et al., 2020; Arul et al., 2006).

5.3 Machining process control through MCDM/ algorithm approach

In optimizing different process constraints, distinguished scholars estimate the optimal parametric conditions for efficient machining. The first and most crucial step for process control and optimization is predictive

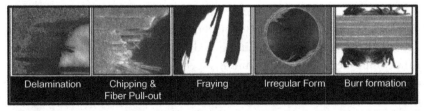

Fig. 5.4 Drilling-induced defects. *From Meinhard, D., Haeger, A., & Knoblauch, V. (2021). Drilling-induced defects on carbon fiber-reinforced thermoplastic polyamide and their effect on mechanical properties. Composite Structures, 256, 113138. https://doi.org/10. 1016/j.compstruct.2020.113138.*

modeling of machining operations. A predictive model represents the relationship between input process parameters and output response. The principal objective of modeling processes must be that the output of machining operations can be reliably predicted quantitatively. Modeling may help with efficient machining process planning for maximum efficiency, quality, and minimum cost. The lack of basic understanding of tooltip and work material interaction is attributed to the lack of machining process modeling (van Luttervelt et al., 1998). There are two steps to predictive modeling of machining processes for practical applications. The first step is developing predictive models for machining parameters, followed by the establishing models for machining output. In stage 1, specific phenomena are expected in chip forming, including stresses, temperatures, friction, contact time of tool chips, chip flow, etc. In stage 2, the machining performance measures needed for the practical application are prediction, such as cutting forces, torque, strength, tool wear/tool life, chip form/chapping, surface roughness, etc. The workpiece properties, shape, and size significantly alter the machining performance. The machining tool material, design, geometry, and lubricants also affect the machining characteristics. Many methodologies can be used to establish the relationships among process inputs and outputs. There are minimal studies that can perform the above-defined task reliably due to the highly non-linear nature of metal cutting operations. The modeling of varying constraints is mainly performed in three stages: analytical, numerical, and empirical. According to van Luttervelt et al. (1998), 43% of research groups were functioning in empirical modeling, 32% in analytical modeling, and 18% in numerical modeling.

5.4 Drilling of laminated polymer nanocomposite: A case study

From the last two decades, eminent scholars performed outstanding work in polymers machining. But current information shows that very limited work is offered on machining and modeling of laminated polymer nanocomposites and their machining aspects. The imprecise surface quality of the machined hole can lead to an inefficient assembly. Also, it could hamper the product strength, efficiency, and stability after a specific time (Brinksmeier, 1990; Saoudi et al., 2016; Sorrentino et al., 2017; Zitoune et al., 2016). In the manufacturing and engineering sectors, 40% of the assembly and production time is dedicated to machining the holes, confirming the many efforts achieved by distinguished researchers (El-Sonbaty et al., 2004; Saoudi et al., 2016; Zitoune et al., 2016).

During machining, the laminate nanocomposites, controllable factors (spindle speed, feed rate, and weight % of nanomaterials) play a significant role in the quality of the developed components and machining efficiency. Polymer machining becomes a complex task due to intermittent cutting force, varying process constraints, and material quality. Composite damage is a phenomenon of interplay/layers failure, a tendency caused by the perforation of composite laminates, which is extremely severe and sometimes causes unpredictable damages and assembly failure in a significant way (Eneyew & Ramulu, 2014; Gemi et al., 2019). This chapter selects a case study for different process parameters to optimize the response value roughness and circularity error. Table 5.1 demonstrates the selection of process parameters and range of constraints. Fig. 5.5 reveals the experiment number corresponding response value.

5.4.1 Mathematical model

Various process parameters were studied to identify their impact on drilling performances after the drilling of polymer nanocomposites. A 95% confidence level has been used to examine the level of influence on machining performance using an ANOVA test (Ajith Arul Daniel et al., 2019). The Fisher test (F-value) was performed to verify the parameter setting (Liu et al., 2010). On the other hand, the P-value implies the importance of factors. Nonlinearity is highly sought in the R^2 value, which can be found in the correlation coefficient of the multiregression coefficients through a nonlinear model. In the context of statistical analysis, if the P-value is less than 0.05, then at least one of the terms in the model is significant in influencing the mean response of the machining output. In further investigation, the model adequacy was found to be adequate. The ANOVA results revealed

Table 5.1 Selection of process parameters and levels.

Milling constraints/unit	Levels		
Spindle speed (rpm)	800	1600	2400
Feed rate (mm/min)	80	160	240
GO (wt%)	1	2	3

From Kumar, J., & Kumar Verma, R. (2021). A novel methodology of Combined Compromise Solution and Principal Component Analysis (CoCoSo-PCA) for machinability investigation of graphene nanocomposites. *CIRP Journal of Manufacturing Science and Technology, 33*, 143–157. https://doi.org/10.1016/j.cirpj.2021.03.007.

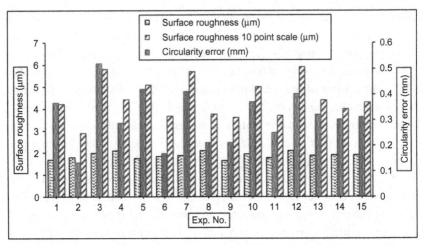

Fig. 5.5 Box Behnken design and response. *From Kumar, J., & Kumar Verma, R. (2021). A novel methodology of combined compromise solution and principal component analysis (CoCoSo-PCA) for machinability investigation of graphene nanocomposites. CIRP Journal of Manufacturing Science and Technology, 33, 143–157. https://doi.org/10.1016/j.cirpj. 2021.03.007.*

that the ideal location for this investigation might be predicted. A wide variety of experimental parameters were factored into the experimentation, which led to the development of an experimental design that incorporates a systematic experimental array. To perform mathematical modeling and evaluate the significant parameters, ANOVA was employed. The method is used to investigate various parameters and their effects on responses in controlled experiments. A statistical technique is utilized to find variances in the actual performance of the evaluation module (Premnath et al., 2014). From Fig. 5.6, it has found the model adequate for the selected response surface roughness indices (R_a) 98.49%, ten-point scale (R_z) 93.93%, and circularity error (C_e) 96.96%.

$$R_a = 1.2897 + 0.000083 \times N + 0.001891 \times f + 0.08000 \times G\% \quad (5.1)$$

$$R_z = 3.829 - 0.000944 \times N + 0.01052 \times f + 0.2137 \times G\% \quad (5.2)$$

$$Ce = 0.3614 - 0.000143 \times N + 0.000966 \times f + 0.01263 \times G\% \quad (5.3)$$

5.4.2 Antlion optimization (ALO) algorithm

Antlions are very small insects available in the holes and known to hunt in two main phases: the larvae and the adult. During the larvae stage, they mainly hunt rodents and invertebrates. They use sit-and-wait tactics for

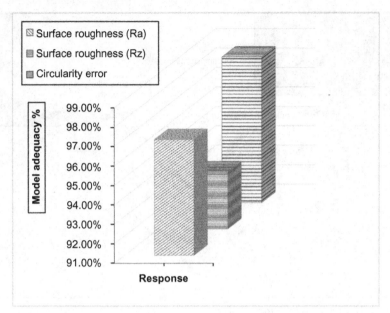

Fig. 5.6 Model adequacy. *From Kumar, J., & Kumar Verma, R. (2021). A novel methodology of combined compromise solution and principal component analysis (CoCoSo-PCA) for machinability investigation of graphene nanocomposites.* CIRP Journal of Manufacturing Science and Technology, 33, *143–157. https://doi.org/10.1016/j.cirpj.2021.03.007.*

hunting their prey. These insect species use a cone-shaped pit to catch insects. After digging the hole, the larvae emerge from the sand and hide underneath the cone base. The cone-shaped traps used by antlions are designed to mimic the movements of ants in the nest. In the ALO, antlions are required to move over the nest to catch the ants. Antlions are different from bug species. During the larvae stage, they hunt small insects like ants. They create a tight trap with a sharp edge on the bottom part of the sand bar, as demonstrated in Fig. 5.7. As the larvae dig, they will periodically dump sand at the bottom of the pit. This study analyzed the performance of the ALO algorithm and results reveal that the algorithm consistently performs better results (Mirjalili, 2015).

It involves establishing the ideal location of the antlion for the prey. The ideal location is then computed and assigned to the antlion. The fittest antlion is then chosen to get the ideal fitness value. The algorithm then tries to get the correct location of the revised generation. If the obtained value is not satisfactory, then the previous final value is considered the final one. The ultimate elite refers to the location of the most practical solution. It can be

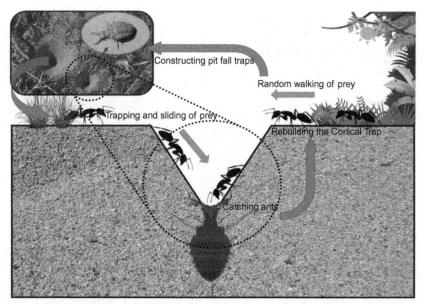

Fig. 5.7 Hunting mechanism of antlion.

obtained by taking the antlions foraging behavior into account. The working of the ALO algorithm is discussed as follows:

Step 1: Random initialization of the first population of ants and antlions. An initial population of ants as $X_{ant} = (x_1, x_2, x_3, \ldots\ldots x_n)$ and antlions as $X_{antlion} = (x_1^*, x_2^*, x_3^*, \ldots\ldots x_n^*)$ is generated within the search space of the parameters, where n is the size of the population.

Step 2: Modeling of random motion of ants.

The ants walk randomly during their food search. The random motion of ants is mathematically modeled as,

$$X(t^{ALO}) = \left[0, Cum_{sum}(2r(t_1^{ALO}) - 1), Cum_{sum}(2r(t_2^{ALO}) - 1), \cdots, \right.$$

$$\left. Cum_{sum}(2r(t_{max\_iter}^{ALO}) - 1)\right]$$

$$(5.4)$$

where Cum_{sum} denotes the cumulative sum, max_iter shows the maximum number of iterations, t^{ant} illustrates the movement of random walk per iteration, and $r(t^{ant})$ is a stochastic function identified as follows.

$$r(t^{ant}) = \begin{cases} 0, & \text{if } rand() \leq 0.5 \\ 1, & \text{if } rand() > 0.5 \end{cases},$$

$$rand() \in [0, 1] \qquad (5.5)$$

The irregular form of random walk of ants is employed to hold the location of ants in the search region.

$$X_i^{t^{ALO}} = \frac{\left(X_i^{t^{ALO}} - a_i\right) \times \left(c_i^{t^{ALO}} - d_i^{t^{ALO}}\right)}{(a_i - b_i)} + c_i^{t^{ALO}} \qquad (5.6)$$

where a, b, c, and d depict the minimum of a random movement, the highest value of a random move, lower and upper boundaries for fitness function, respectively.

An antlion pit-based trap in the ALO algorithm is constructed by utilizing a roulette wheel to determine the antlions with the highest fitness value. This system encourages a great likelihood to capture prey.

Step 3: Trapping and sliding of ants in antlion pit can be mathematically modeled as,

Trap construction model:

$$c_i^{t^{ALO}} = x_{Antlion_j}^{t^{ALO}} + c^{t^{ALO}} \qquad (5.7)$$

$$d_i^{t^{ant}} = x_{Antlion_j}^{t^{ALO}} + d^{t^{ALO}} \qquad (5.8)$$

where $Antlion_j$ displays the location of the jth Antlion at iteration any t^{ALO} iteration.

Ants slipping toward antlion: The antlion throws sand near the edge of the trap to lure the ant into crawling into the trap when the ant attempts to crawl away.

$$c^{t^{ALO}} = \frac{c^{t^{ALO}}}{I} \qquad (5.9)$$

$$d^{t^{ALO}} = \frac{d^{t^{ALO}}}{I} \qquad (5.10)$$

where

$$I = 10^w \times \frac{t^{ALO}}{\max\_iter},$$

$$w = \begin{cases} 2, & \text{when } t^{ALO} > 0.1 \text{ of } max\_iter \\ 3, & \text{when } t^{ALO} > 0.5 \text{ of } max\_iter \\ 4, & \text{when } t^{ALO} > 0.75 \text{ of } max\_iter \\ 5, & \text{when } t^{ALO} > 0.9 \text{ of } max\_iter \\ 6, & \text{when } t^{ALO} > 0.95 \text{ of } max\_iter \end{cases}$$

Step 4: Updating the antlion position.

If the fitness function value of ant is greater than the antlion, the ant's position is allotted to antlion.

$$Antlion_j^{t^{ALO}} = Ant_j^{t^{ALO}} \, if f\left(Ant_j^{t^{ALO}}\right) < f\left(Antlion_j^{t^{ALO}}\right) \tag{5.11}$$

Step 5: Updating the elitism.

The best antlion picked at each iteration is the elite solution. The elite was held the fittest antlion, and it influenced all the movements of the ants, while the iterative runs. It is believed that the ants are walking randomly around the chosen antlion as per the roulette wheel controller. Antlion elitism can be measured as:

$$Ant_j^{t^{ALO}} = \frac{R_A^{t^{ALO}} + R_E^{t^{ALO}}}{2} \tag{5.12}$$

where $R_A^{t^{ALO}}$ represents a random movement across the chosen antlion on a roulette wheel, and $R_E^{t^{ALO}}$ shows the elite value at t^{ALO} iteration, respectively.

Step 7: Repeat the above steps until the termination condition is assured.

Pseudocode for ALO

Begin

 Initialization of First Population of prey and antlions with a random solution.

 Evaluate the fitness value for all prey and antlions.

 Find out the best Antlion and set it as elite value (Determination of optimum value).

 While the termination criterion is not fulfilled.

  for every prey

  Random selection of Antlion using Roulette Wheel function.

  Update values of c and d using Eqs. (5.7) and (5.8).

  Definition of random walk of prey around the pit and normalized using Eqs. (5.4)–(5.6).

Update location value of prey (ant/bug) using Eq. (5.12).
end for
Update the fitness value of every agent (prey/antlion).
Substitute antlion with its consequent prey (ant/bug) using Eq. (5.11)
Update elite if an antlion becomes fitter than elite.
end While
Return fitness value (elite).
End

During the drilling experimentation with the spindle speed of 1600 rpm, feed rate of 80 mm/min, and weight percentage of graphene oxide at 1%, the lower value of surface roughness (R_a) is observed. In the surface roughness evaluation, the Ra value of 1.80 μm is detected, which is lower than the obtained value in the assessed experimental array. Similar observations have been for the surface roughness (R_z) and circularity error (C_e), which are evaluated to be 2.9 μm and 0.133, respectively, with parametric settings of the spindle speed being 2400 rpm, the feed rate being 80 mm/min, and the percentage of graphene oxide being 2%. Drilling is a more complicated process since it includes an intermittent cutting function. Therefore, the temperature should be high enough to generate deformation in the plastic deformation region during the drilling process. However, the temperature should not be too high that the cutting tool features negatively affect the machining process, ensuring the need to soften the cutting zone for easier machinability. The feed rate can be maintained at 80 mm/min to regulate the situation mentioned overhead. Concerning the hole damage (circularity error), it is seen that the slightest degree of circularity error is achieved at a lower feed rate of 80 mm/min. Also, the weight % of GO in the fiber breaks goes down; cracking and breaking due to an increase in the laminar strength of the polymer composite occurs (Jenarthanan & Jeyapaul, 2013).

Fig. 5.8 depicts the ALO algorithm predicted value convergence graph. The outcome of input parameters over the ALO algorithm is a spindle speed of 800 rpm, a feed rate of 80 mm/min, and a wt% of graphene oxide of 1%, resulting in the lowest machined surface indices $(R_a$ and $R_z)$ and circularity error (Ce). For common values of surface roughness and circularity error, a low feed rate is favored. Because at a higher feed rate, the force of the drilling push on the work sample is more remarkable, resulting in more significant flexural that can lead to crack formation (Jenarthanan & Jeyapaul, 2013; Köklü et al., 2019; Kyratsis et al., 2018; Wang et al., 2004). A similar result and machining performance trend were noted in earlier investigations performed by distinguished researchers (Geier & Szalay, 2017; Ragunath et al., 2017).

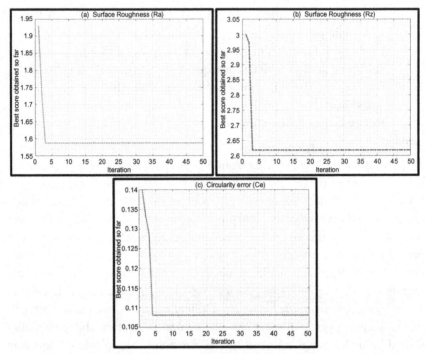

Fig. 5.8 Convergence plot for process response (A) for R_a, (B) for R_z, and (C) for C_e.

The proposed polymer laminates nanocomposite is highly prone to drilling-induced damages such as delamination, matrix debonding, and fiber breakage. It can be controlled by the selection of proper parameters and the adding of GO wt%.

The comparative analysis was done at the optimal condition of the present study over the previous survey (Table 5.2). The outcome from this study has been found that drilling response has significantly improved. Surface roughness indices (R_a) and (R_z) have diminished by 11.66% and 10%,

Table 5.2 Comparative analysis.

	Present study				Previous study
	Array		ALO algorithm		
Response	Value	Settings	Value	Settings	Kumar et al.
R_a	1.80	Exp. No. 9	1.59	N:1 f:1 G%:1	1.72
R_z	2.90	Exp. No. 2	2.61	N:3 f:1 G%:1	2.61
C_e	0.133	Exp. No. 2	0.108	N:3 f:1 G%:1	0.108

respectively. The circularity error (C_e) also decreased by 18.79%. The lower feed rate significantly influenced an enhanced surface finishing by reducing vibration and defects and cracking. In addition, the filler nanomaterial improves the bonding strength, which is critical in the machining test condition.

5.5 Summary of chapter

This chapter contains the machining investigations and parameter modeling while drilling of modified epoxy nanocomposites. The control of process parameters and their influence on the reduction of damages and defects has been explored. The impact of nanofiller on machining and ALO algorithm used to investigate drilling machinability is discussed in this chapter. The proposed nanocomposites are found capable of creating different holes through drilling operations. The ALO algorithm has evaluated the optimal parametric conditions. Based on the study findings, the following conclusion can be made:

The previous study was intended for machining polymer composite as a limited investigation was focusing on the design of the experiment. It consumes lots of time taken, is costly, inaccurate, errors in the results, etc.

ANOVA results for surface roughness indices (R_a and R_z) and circularity error demonstrate that the developed model is R_a-98.49%, R_z-93.93%, and C_e-96.96% adequacy, which is remarked as satisfactory. Later, the fitness function is generated to optimize the response using an intelligence algorithm approach, which is extremely difficult in conventional optimization tools.

The optimal setting through the ALO algorithm is found as a practical approach confirmed by a previous study that reveals acceptable.

The algorithm can be recommended to the different engineering processes and multicriteria optimization case studies.

Micromanufacturing is another emerging area for practitioners and academia, as it can be used in electronics devices, optical components, aviation parts, etc. Hence, the developed material and optimization tools can be explored for the microlevel machining process. The current study may be summarized by claiming that the proposed optimization module can be recommended for critical multicriteria decision-making (MCDM) problems in manufacturing and industrial engineering case studies. The proposed optimization tool can be customized for other machining procedures (turning, broaching, grinding) of the developed laminated polymer nanocomposites.

The other machining constraints, like tool material, chip pattern mechanics, material removal, tooltip temperature, tool wear, etc., may be considered to interpret the polymer nanocomposite machining features. The mathematical models give a satisfactory agreement in results. It can also approximate and predict manufacturing procedures such as grinding, turning, die sinking, welding, etc., and other complex case studies of industrial engineering.

Acknowledgments

The authors are very thankful to the Uttar Pradesh Council of Science and Technology, Lucknow, India.

Funding

This work is financially supported under the R & D project (ID-CSTD-2491) scheme of the Uttar Pradesh Council of Science and Technology, Lucknow, India.

Declaration of conflicting interests

The authors declared no potential conflicts of interest.

References

Aamir, M., Giasin, K., Tolouei-Rad, M., & Vafadar, A. (2020). A review: Drilling performance and hole quality of aluminium alloys for aerospace applications. *Journal of Materials Research and Technology*, *9*(6), 12484–12500. https://doi.org/10.1016/j.jmrt.2020.09.003.

Adak, N. C., Chhetri, S., Sabarad, S., Roy, H., Murmu, N. C., Samanta, P., et al. (2019). Direct observation of micro delamination in graphene oxide incorporated carbon fiber/epoxy composite via in-situ tensile test. *Composites Science and Technology*, *177*, 57–65. https://doi.org/10.1016/j.compscitech.2019.04.006.

Ajith Arul Daniel, S., Pugazhenthi, R., Kumar, R., & Vijayananth, S. (2019). Multi objective prediction and optimization of control parameters in the milling of aluminium hybrid metal matrix composites using ANN and Taguchi -grey relational analysis. *Defence Technology*, *15*(4), 545–556. https://doi.org/10.1016/j.dt.2019.01.001.

Allwin Roy, Y., Gobivel, K., Vijay Sekar, K. S., & Suresh Kumar, S. (2017). Impact of cutting forces and chip microstructure in high speed machining of carbon fiber—Epoxy composite tube. *Archives of Metallurgy and Materials*, *62*(3). http://journals.pan.pl/dlibra/publication/edition/105128.

Anarghya, A., Harshith, D. N., Rao, N., Nayak, N. S., Gurumurthy, B. M., Abhishek, V. N., et al. (2018). Thrust and torque force analysis in the drilling of aramid fibre-reinforced composite laminates using RSM and MLPNN-GA. *Heliyon*, *4*(7). https://doi.org/10.1016/j.heliyon.2018.e00703, e00703.

Anh Nguyen, T., Tung Nguyen, Q., & Phuc Bach, T. (2019). Mechanical properties and flame retardancy of epoxy resin/nanoclay/multiwalled carbon nanotube nanocomposites. *Journal of Chemistry*, *2019*, 3105205 (1-9) https://doi.org/10.1155/2019/3105205.

Arul, S., Vijayaraghavan, L., Malhotra, S. K., & Krishnamurthy, R. (2006). The effect of vibratory drilling on hole quality in polymeric composites. *International Journal of*

Machine Tools and Manufacture, *46*(3), 252–259. https://doi.org/10.1016/j.ijmachtools. 2005.05.023.

Ashby, M. F., & Ashby, M. F. (2013). *Material selection strategies* (pp. 227–273). Butterworth-Heinemann. https://doi.org/10.1016/B978-0-12-385971-6.00009-9. Chapter 9.

Ashby, M. F., Bréchet, Y. J. M., Cebon, D., & Salvo, L. (2004). Selection strategies for materials and processes. *Materials & Design*, *25*(1), 51–67. https://doi.org/10.1016/S0261-3069(03)00159-6.

Bal, S., & Saha, S. (2014). Comparison and analysis of physical properties of carbon nanomaterial-doped polymer composites. *High Performance Polymers*, *26*(8), 953–960. https://doi.org/10.1177/0954008314535823.

Bhatnagar, N., Singh, I., & Nayak, D. (2004). Damage investigation in drilling of glass fiber reinforced plastic composite laminates. *Null*, *19*(6), 995–1007. https://doi.org/10.1081/AMP-200034486.

Bielawski, R., Kowalik, M., Suprynowicz, K., Rządkowski, W., & Pyrzanowski, P. (2017). Experimental study on the riveted joints in glass fibre reinforced plastics (GFRP). *Archive of Mechanical Engineering*, *64*(3), 301–313. http://journals.pan.pl/dlibra/publication/edition/104243.

Brebu, M. (2020). Environmental degradation of plastic composites with natural fillers—A review. *Polymers*, *12*(1). https://doi.org/10.3390/polym12010166.

Brinksmeier, E. (1990). Prediction of tool fracture in drilling. *CIRP Annals*, *39*(1), 97–100. https://doi.org/10.1016/S0007-8506(07)61011-7.

Burkov, M., & Eremin, A. (2018). Hybrid CFRP/SWCNT composites with enhanced electrical conductivity and mechanical properties. *Journal of Materials Engineering and Performance*, *27*(11), 5984–5991. https://doi.org/10.1007/s11665-018-3695-x.

Cao, K., Feng, S., Han, Y., Gao, L., Hue Ly, T., Xu, Z., et al. (2020). Elastic straining of free-standing monolayer graphene. *Nature Communications*, *11*(1), 284. https://doi.org/10.1038/s41467-019-14130-0.

Changgu, L., Xiaoding, W., Kysar, J. W., & James, H. (2008). Measurement of the elastic properties and intrinsic strength of monolayer graphene. *Science*, *321*(5887), 385–388. https://doi.org/10.1126/science.1157996.

Cui, M., Ren, S., Pu, J., Wang, Y., Zhao, H., & Wang, L. (2019). Poly(o-phenylenediamine) modified graphene toward the reinforcement in corrosion protection of epoxy coatings. *Corrosion Science*, *159*. https://doi.org/10.1016/j.corsci.2019.108131, 108131.

Dahnel, A. N., Ascroft, H., & Barnes, S. (2016). The effect of varying cutting speeds on tool Wear during conventional and ultrasonic assisted drilling (UAD) of carbon fibre composite (CFC) and titanium alloy stacks. In *Vol. 46. 7th HPC 2016—CIRP conference on high performance cutting* (pp. 420–423). https://doi.org/10.1016/j.procir.2016.04.044.

El-Sonbaty, I., Khashaba, U. A., & Machaly, T. (2004). Factors affecting the machinability of GFR/epoxy composites. *Composite Structures*, *63*(3), 329–338. https://doi.org/10.1016/S0263-8223(03)00181-8.

Eneyew, E. D., & Ramulu, M. (2014). Experimental study of surface quality and damage when drilling unidirectional CFRP composites. *Journal of Materials Research and Technology*, *3*(4), 354–362. https://doi.org/10.1016/j.jmrt.2014.10.003.

Findik, F., & Turan, K. (2012). Materials selection for lighter wagon design with a weighted property index method. *Materials & Design*, *37*, 470–477. https://doi.org/10.1016/j.matdes.2012.01.016.

Fleischer, J., Teti, R., Lanza, G., Mativenga, P., Möhring, H.-C., & Caggiano, A. (2018). Composite materials parts manufacturing. *CIRP Annals*, *67*(2), 603–626. https://doi.org/10.1016/j.cirp.2018.05.005.

Gang, D. (2017). The effect of surface treatment of CF and graphene oxide on the mechanical properties of PI composite. *Journal of Thermoplastic Composite Materials*, *31*(9), 1219–1231. https://doi.org/10.1177/0892705717734606.

Gara, S., M'hamed, S., & Tsoumarev, O. (2018). Temperature measurement and machining damage in slotting of multidirectional CFRP laminate. *Null*, *22*(2), 320–337. https://doi.org/10.1080/10910344.2017.1365892.

Gaugel, S., Sripathy, P., Haeger, A., Meinhard, D., Bernthaler, T., Lissek, F., et al. (2016). A comparative study on tool wear and laminate damage in drilling of carbon-fiber reinforced polymers (CFRP). *Composite Structures*, *155*, 173–183. https://doi.org/10.1016/j.compstruct.2016.08.004.

Geier, N., & Szalay, T. (2017). Optimisation of process parameters for the orbital and conventional drilling of uni-directional carbon fibre-reinforced polymers (UD-CFRP). *Measurement*, *110*, 319–334. https://doi.org/10.1016/j.measurement.2017.07.007.

Gemi, L., Morkavuk, S., Köklü, U., & Gemi, D. S. (2019). An experimental study on the effects of various drill types on drilling performance of GFRP composite pipes and damage formation. *Composites Part B: Engineering*, *172*, 186–194. https://doi.org/10.1016/j.compositesb.2019.05.023.

George, J., Sreekala, M. S., & Thomas, S. (2004). A review on interface modification and characterization of natural fiber reinforced plastic composites. *Polymer Engineering and Science*, *41*(9), 1471–1485. https://doi.org/10.1002/pen.10846.

Giasin, K., Hodzic, A., Phadnis, V., & Ayvar-Soberanis, S. (2016). Assessment of cutting forces and hole quality in drilling Al2024 aluminium alloy: Experimental and finite element study. *The International Journal of Advanced Manufacturing Technology*, *87*(5), 2041–2061. https://doi.org/10.1007/s00170-016-8563-y.

Haeger, A., Meinhard, D., Lissek, F., Kaufeld, M., Hoffmann, M. J., Schneider, G., et al. (2016). Interaction between laminate quality, drilling-induced delamination and mechanical properties in machining of carbon fibre reinforced plastic (CFRP). *Materialwissenschaft und Werkstofftechnik*, *47*(11), 997–1014. https://doi.org/10.1002/mawe.201600626.

Haeger, A., Schoen, G., Lissek, F., Meinhard, D., Kaufeld, M., Schneider, G., et al. (2016). Non-destructive detection of drilling-induced delamination in CFRP and its effect on mechanical properties. In *Vol. 149. International conference on manufacturing engineering and materials, ICMEM 2016, 6–10 June 2016, Nový Smokovec, Slovakia* (pp. 130–142). https://doi.org/10.1016/j.proeng.2016.06.647.

Ho-Cheng, H., & Dharan, C. K. H. (1990). Delamination during drilling in composite laminates. *Journal of Engineering for Industry*, *112*(3), 236–239. https://doi.org/10.1115/1.2899580.

Hocheng, H., & Tsao, C. C. (2003). Comprehensive analysis of delamination in drilling of composite materials with various drill bits. *Journal of Materials Processing Technology*, *140*(1), 335–339. https://doi.org/10.1016/S0924-0136(03)00749-0.

Jenarthanan, M. P., & Jeyapaul, R. (2013). Optimisation of machining parameters on milling of GFRP composites by desirability function analysis using Taguchi method. *International Journal of Engineering, Science and Technology*, *5*(4), 23–36. https://doi.org/10.4314/ijest.v5i4.3.

John, K., Kumaran, S. T., Kurniawan, R., Moon Park, K., & Byeon, J. (2018). Review on the methodologies adopted to minimize the material damages in drilling of carbon fiber reinforced plastic composites. *Journal of Reinforced Plastics and Composites*, *38*(8), 351–368. https://doi.org/10.1177/0731684418819822.

Kato, A., Ikeda, Y., & Kohjiya, S. (2012). Carbon black-filled natural rubber composites: Physical chemistry and reinforcing mechanism. In *Polymer Composites* Willey. https://doi.org/10.1002/9783527645213.ch17.

Khashaba, U. A. (2004). Delamination in drilling GFR-thermoset composites. *Composite Structures*, *63*(3), 313–327. https://doi.org/10.1016/S0263-8223(03)00180-6.

Köklü, U., Mayda, M., Morkavuk, S., Avcı, A., & Demir, O. (2019). Optimization and prediction of thrust force, vibration and delamination in drilling of functionally graded composite using Taguchi, ANOVA and ANN analysis. *Materials Research Express*, 6(8). https://doi.org/10.1088/2053-1591/ab2617, 085335.

Komanduri, R. (1997). Machining of fiber-reinforced composites. *Null*, 1(1), 113–152. https://doi.org/10.1080/10940349708945641.

Kumar, J., & Kumar Verma, R. (2020). Experimental investigations and multiple criteria optimization during milling of graphene oxide (GO) doped epoxy/CFRP composites using TOPSIS-AHP hybrid module. *FME Transactions*, 48, 628–635. https://doi.org/10.5937/fme2003628K.

Kumar, A., Sharma, K., & Dixit, A. R. (2019). A review of the mechanical and thermal properties of graphene and its hybrid polymer nanocomposites for structural applications. *Journal of Materials Science*, 54(8), 5992–6026. https://doi.org/10.1007/s10853-018-03244-3.

Kumar, D., & Singh, K. (2017). Investigation of delamination and surface quality of machined holes in drilling of multiwalled carbon nanotube doped epoxy/carbon fiber reinforced polymer nanocomposite. *Proceedings of the Institution of Mechanical Engineers, Part L: Journal of Materials: Design and Applications*, 233(4), 647–663. https://doi.org/10.1177/1464420717692369.

Kwon, H., Mondal, J., AlOgab, K. A., Sammelselg, V., Takamichi, M., Kawaski, A., et al. (2017). Graphene oxide-reinforced aluminum alloy matrix composite materials fabricated by powder metallurgy. *Journal of Alloys and Compounds*, 698, 807–813. https://doi.org/10.1016/j.jallcom.2016.12.179.

Kyratsis, P., Markopoulos, A. P., Efkolidis, N., Maliagkas, V., & Kakoulis, K. (2018). Prediction of thrust force and cutting torque in drilling based on the response surface methodology. *Machines*, 6(2). https://doi.org/10.3390/machines6020024.

Layth, M., Ansari, M. N., Pua, G., Mohammad, J., & Saiful Islam, M. (2015). A review on natural fiber reinforced polymer composite and its applications. *International Journal of Polymer Science*, 2015, 243947 (1-15) https://doi.org/10.1155/2015/243947.

Liu, Y.-T., Chang, W.-C., & Yamagata, Y. (2010). A study on optimal compensation cutting for an aspheric surface using the Taguchi method. *CIRP Journal of Manufacturing Science and Technology*, 3(1), 40–48. https://doi.org/10.1016/j.cirpj.2010.03.001.

Liu, D., Tang, Y., & Cong, W. L. (2012). A review of mechanical drilling for composite laminates. *Composite Structures*, 94(4), 1265–1279. https://doi.org/10.1016/j.compstruct.2011.11.024.

López, N., de Lacalle, L., Campa, F. J., Lamikiz, A., & Paulo Davim, J. (2011). *Milling* (pp. 213–303). Woodhead Publishing. https://doi.org/10.1533/9780857094940.213.3.

Mirjalili, S. (2015). The ant lion optimizer. *Advances in Engineering Software*, 83, 80–98. https://doi.org/10.1016/j.advengsoft.2015.01.010.

Myagkov, L. L., Mahkamov, K., Chainov, N. D., Makhkamova, I., & Folkson, R. (2014). *Advanced and conventional internal combustion engine materials* (pp. 370–408e). Woodhead Publishing. https://doi.org/10.1533/9780857097422.2.370.11.

Nijssen, R. P. L. (2015). *Composite materials: An introduction*. Inholland University of Applied Sciences.

Ozkan, D., Sabri Gok, M., Oge, M., & Cahit Karaoglanli, A. (2019). Milling behavior analysis of carbon fiber-reinforced polymer (CFRP) composites. *International Conference on Modern Trends in Manufacturing Technologies and Equipment*, 2018(11), 526–533. https://doi.org/10.1016/j.matpr.2019.01.024.

Papageorgiou, D. G., Liu, M., Li, Z., Vallés, C., Young, R. J., & Kinloch, I. A. (2019). Hybrid poly(ether ether ketone) composites reinforced with a combination of carbon fibres and graphene nanoplatelets. *Composites Science and Technology*, 175, 60–68. https://doi.org/10.1016/j.compscitech.2019.03.006.

Pathak, A. K., Borah, M., Gupta, A., Yokozeki, T., & Dhakate, S. R. (2016). Improved mechanical properties of carbon fiber/graphene oxide-epoxy hybrid composites.

Composites Science and Technology, *135*, 28–38. https://doi.org/10.1016/j.compscitech. 2016.09.007.

Premnath, A. A., Alwarsamy, T., & Sugapriya, K. (2014). A comparative analysis of tool wear prediction using response surface methodology and artificial neural networks. *Null*, *12* (1), 38–48. https://doi.org/10.7158/M12-075.2014.12.1.

Ragunath, S., Velmurugan, C., & Kannan, T. (2017). Optimization of drilling delamination behavior of GFRP/clay nano-composites using RSM and GRA methods. *Fibers and Polymers*, *18*(12), 2400–2409. https://doi.org/10.1007/s12221-017-7420-4.

Ray, S., & Easteal, A. J. (2007). Advances in polymer-filler composites: Macro to nano. *Materials and Manufacturing Processes*, *22*(6), 741–749. https://doi.org/10.1080/ 10426910701385366.

Saadatmandi, S., Ramezanzadeh, B., Asghari, M., & Bahlakeh, G. (2020). Graphene oxide nanoplatform surface decoration by spherical zinc-polypyrrole nanoparticles for epoxy coating properties enhancement: Detailed explorations from integrated experimental and electronic-scale quantum mechanics approaches. *Journal of Alloys and Compounds*, *816*. https://doi.org/10.1016/j.jallcom.2019.152510, 152510.

Santhanakrishnan Balakrishnan, V., Seidlitz, H., Yellur, M. R., & Vogt, N. (2019). A study on the influence of drilling and CO2 laser cutting in carbon/epoxy laminates. *Journal of Materials Research and Technology*, *8*(1), 944–949. https://doi.org/10.1016/j. jmrt.2018.05.025.

Saoudi, J., Zitoune, R., Mezlini, S., Gururaja, S., & Seitier, P. (2016). Critical thrust force predictions during drilling: Analytical modeling and X-ray tomography quantification. *Composite Structures*, *153*, 886–894. https://doi.org/10.1016/j.compstruct.2016.07.015.

Sengupta, R., Bhattacharya, M., Bandyopadhyay, S., & Bhowmick, A. K. (2011). A review on the mechanical and electrical properties of graphite and modified graphite reinforced polymer composites. *Special Issue on Conducting Polymers*, *36*(5), 638–670. https://doi. org/10.1016/j.progpolymsci.2010.11.003.

Slamani, M., Gauthier, S., & Chatelain, J.-F. (2015). A study of the combined effects of machining parameters on cutting force components during high speed robotic trimming of CFRPs. *Measurement*, *59*, 268–283. https://doi.org/10.1016/j.measurement. 2014.09.052.

Sorrentino, L., Turchetta, S., & Bellini, C. (2017). In process monitoring of cutting temperature during the drilling of FRP laminate. *Composite Structures*, *168*, 549–561. https:// doi.org/10.1016/j.compstruct.2017.02.079.

Stewart, R. (2009). Lightweighting the automotive market. *Reinforced Plastics*, *53*(2), 14–21. https://doi.org/10.1016/S0034-3617(09)70078-5.

Sun, F., & Zhao, Y.-P. (2020). Geomaterials evaluation: A new application of Ashby plots. *Materials (Basel, Switzerland)*, *13*(11), 2517. https://doi.org/10.3390/ma13112517.

Uygunoglu, T., Gunes, I., & Brostow, W. (2015). Physical and mechanical properties of polymer composites with high content of wastes including boron. *Materials Research*, *15*(6), 1188–1196. https://doi.org/10.1590/1516-1439.009815.

van Luttervelt, C. A., Childs, T. H. C., Jawahir, I. S., Klocke, F., Venuvinod, P. K., Altintas, Y., et al. (1998). Present situation and future trends in modelling of machining operations progress report of the CIRP working group 'modelling of machining operations'. *CIRP Annals*, *47*(2), 587–626. https://doi.org/10.1016/S0007-8506(07)63244-2.

Voß, R., Henerichs, M., Kuster, F., & Wegener, K. (2014). Chip root analysis after machining carbon fiber reinforced plastics (CFRP) at different Fiber orientations. In *Vol. 14. 6th CIRP international conference on high performance cutting, HPC2014* (pp. 217–222). https:// doi.org/10.1016/j.procir.2014.03.013.

Voss, R., Seeholzer, L., Kuster, F., & Wegener, K. (2017). Influence of fibre orientation, tool geometry and process parameters on surface quality in milling of CFRP. *CIRP Journal of Manufacturing Science and Technology*, *18*, 75–91. https://doi.org/10.1016/j.cirpj.2016.10.002.

Wang, X., Wang, L. J., & Tao, J. P. (2004). Investigation on thrust in vibration drilling of fiber-reinforced plastics. *Journal of Materials Processing Technology*, *148*(2), 239–244. https://doi.org/10.1016/j.jmatprotec.2003.12.019.

Wang, R.-M., Zheng, S.-R., Zheng, Y.-P., Wang, R.-M., Zheng, S.-R., & Zheng, Y.-P. (2011). Introduction to polymer matrix composites. In *Woodhead publishing series in composites science and engineering* (pp. 1–548). Woodhead Publishing. https://doi.org/10.1533/9780857092229.1. 1.

Zitoune, R., Krishnaraj, V., Collombet, F., & Le Roux, S. (2016). Experimental and numerical analysis on drilling of carbon fibre reinforced plastic and aluminium stacks. *Composite Structures*, *146*, 148–158. https://doi.org/10.1016/j.compstruct.2016.02.084.

CHAPTER SIX

Machining performance analysis of micro-ED milling process of titanium alloy (Ti-6Al-4V)

P. Sivaprakasam[a], P. Hariharan[b], S. Gowri[b], and J. Udaya Prakash[c]

[a]Department of Mechanical Engineering, College of Electrical and Mechanical Engineering, Center of Excellence - Nano Technology, Addis Ababa Science and Technology University, Addis Ababa, Ethiopia
[b]Department of Manufacturing Engineering, College of Engineering Guindy, Anna University, Chennai, India
[c]Department of Mechanical Engineering, Vel Tech Rangarajan Dr. Sagunthala R & D Institute of Science and Technology, Chennai, India

6.1 Introduction

Titanium alloy is broadly applied in chemical plants, power generation, automobile, biomedical applications, and aerospace, due to its appealing qualities, including high thermal strength, low density, high wear and corrosion resistance, low elastic modulus, and biocompatibility. Because of its high hardness and other qualities, titanium alloy can take the challenge of cutting materials, resulting in higher machining forces, reduced MRR, high cutting tool wear, tool vibration, and poor surface finish by conventional machining (Pramanik, Basak, & Prakash, 2019). In comparison with traditional machining, EDM is more advantageous for machining this alloy. Due to recent advances in the industry, the need for microparts and molds has grown rapidly. As a result, advanced manufacturing procedures are ideal for processing these materials.

Product miniaturization depends on various elements for the possible technical development of a variety of high-precision products in industries such as aircraft, environment, biomedical, automotive, electronics, and so on. It has a very high surface-to-volume ratio and can give smaller footprints, energy efficiency, and improved heat transfer. Traditional manufacturing methods like turning, drilling, and milling could be used to machine microfeatures. Still, these machining methods have limitations of excessive tool wear, high aspect ratio, complex shape, and difficult-to-machine (hard) materials like tool steel, super alloy, composites, etc. In these techniques, micro-EDM is an effective technique favored for better surface integrity, high aspect ratio, and low cost due to the contact less method, irrespective of materials hardness (Qin, 2015). Micro-ED milling is used mainly in the micro-manufacturing sector for hard-to-cut materials using cylindrical micro-electrodes. The complex form could be easily machined with specified paths while tool rotating like traditional milling. The removal of material is layer by layer, and the thickness of the layer ranges from a few μm to 0.1 mm (Bissacco, Hansen, Tristo, & Valentincic, 2011).

At present, micro-ED milling is successfully applied to machining 3D cavities, microchannels, complex shapes, etc. (Huang, Bai, Lu, & Guo, 2009). Micro-ED milling, unlike die sinking, does not require a structural tool to create a 3D cavity (Bleys, Kruth, & Lauwers, 2004; Nakagawa, Yuzawa, Sampei, & Hirata, 2017). Hence, it could easily machine hard materials with intricate shapes and sharp corners (Narasimhan, Yu, & Rajurkar, 2005; Yu, Masuzawa, & Fujino, 1998). It also has important challenges to achieve dimensional accuracy for depth and width for a particular fixed length (Ali, Rahman, Zuhaida Zunairi, & Banu, 2017).

Micro-ED milling has a higher TWR than micro-ED drilling because of tool movement in horizontal and vertical directions. The micro-ED milling method is dependent on the layer depth and feed rate for stable machining. Lower depth of cut and feed rate are preferred in the micro-ED milling process, reducing process efficiency (Cheng, Wang, Kobayashi, Nakamoto, & Yamazaki, 2009). In micro-ED milling, thin layers of material are removed one after the other. TW is offset at the end of each layer by normal feeding of the device in the Z-axis. Thin-layer machining wear occurs only at the end of the tool, thus preventing damage to the lateral surfaces. Concise wear measurement is vital since an error in the assessment would have a combined effect in the following layers (Pham, Dimov, Bigot, Ivanov, & Popov, 2004). Lower discharge energy results in fewer craters and peaks, resulting in a smoother surface with a bigger fractal dimension.

To achieve smooth chip evacuation, the tool electrode rotation speed should be kept to a minimum. These result in delicate surface morphology and a large fractal dimension. Coarse surfaces with small fractal dimensions are formed at both high and low gap voltages (Feng, Chu, Hong, & Zhang, 2018). D'Urso et al. (2020) employed an adaptive control strategy for reducing cavity depth error in micro-ED milling. For this specific case, there is a 27% productivity improvement.

Surface morphology and machining quality are the two most important factors in micro-ED milling (Kuriachen & Mathew, 2016). Moses and Jahan conducted a number of studies to compare the microscale properties and surface morphology of Ti–6Al–4V with brass (Moses & Jahan, 2015).

Koyano, Sugata, Hosokawa, and Furumoto (2017) explored less energy to increase the quality of stainless steel surfaces machined. Rajurkar and Lin (2011) conducted a comprehensive review of EDM on titanium alloy and the parameters that influence it. The impact of process variables, including dielectric properties on machining stability, machining performance, and surface integrity, was also investigated to find the critical process variable influencing MRR and TWR. Meena and Azad (2012) used micro-EDM to model and optimize the material removal. A neural system was developed to estimate MRR, and a genetic algorithm was used to optimize machining parameters. Karthikeyan, Ramkumar, Dhamodaran, and Aravindan (2010) investigated micro-ED milling of EN24 steel experimentally. MRR and TWR were examined to characteristics such as speed, aspect ratio, feed, and discharge energy. The rotation speed of the tool is critical in removing debris and providing distributing spark energy evenly in the interelectrode gap.

Yeo, Aligiri, Tan, and Zarepour (2009) proposed a new EDM process monitoring approach. Using this procedure, the pulse is identified and classified as discharge conditions with normal and delayed, arcing, or short circuit. The adaptive speed control servo system method was reducing the machining time. The physical behavior of the micro-ED milling was evaluated based on channel geometry, shape, and machined surface topography in a subsequent study (Karthikeyan, Garg, Ramkumar, & Dhamodaran, 2012).

Tool rotation is a significant micro-ED milling process parameter because it impacts the flow of molten metal, flushing of debris particles, and re-depositing. The instrument's motion interrupts the plasma while altering the channel's dimensions (shape and form). The magnetic field and ultrasonic vibration were explored and compared separately and combined for improvement of MRR and TWR by Jafferson, Hariharan, and Ram Kumar (2014). To achieve a high MRR, researchers recommended using it alone. Arun Pillai and Hariharan (2021) attempted to increase the quality of milled surfaces without losing machining features by employing a nanopowder-mixed dielectrics medium. Magnetic field-introduced micro-ED milling of inconel alloy was investigated; higher MRR has found using magnetic field with the same process parameters (Sivaprakasam, Hariharan, & Elias, n.d.).

This research focuses on micro-ED milling process to produce better surface roughness while enhancing MRR and minimizing TWR. It investigates and discusses the effect of factors, including voltage, capacitance, and feed rate, on output responses. The RSM-based design of experiments

(DoE) was employed for statistical models development. The study discusses the application of established models for multiple-response optimization, and experimentation was also looked into for model verification.

6.2 Experimental details-materials and methods

The DT-110 Micro machine tool (Microtools, Singapore) used for micro-ED milling with RC-type circuit is shown in Fig. 6.1.. The titanium alloy sheet was used with a thickness of 1 mm and a size of 50 mm × 10 mm (L × W). The studies were carried out using a 500-µm-diameter tungsten microtool electrode at a rotational speed of 100 rpm. Microchannels of 1 mm were machined with a depth of 100 µm. The dielectric fluid is EDM 3 oil. Table 6.1 presents the input parameters as well as their levels. RSM was used to conduct various experimental combinations based on a central composite design. Table 6.2 displays the experimental data for machining performance of MRR, TWR, and SR. Eqs. (6.1) and (6.2) are used to compute MRR and TWR. SR values were measured with a non-contact 3D profilometer (CCI Lite, Taylor Hobson).

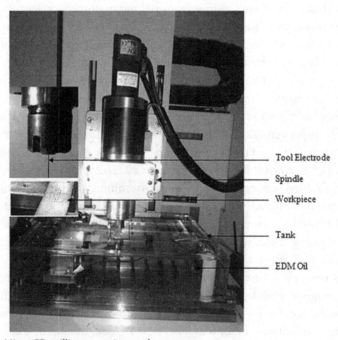

Fig. 6.1 Micro-ED milling experimental setup.

Table 6.1 Factors and levels for micro-ED milling.

	Control factors		
Levels	A: Voltage (V)	B: Capacitance (µF)	C: Feed rate (µm/s)
Low (−1)	100	0.0001	1
Medium (0)	110	0.01	2
High (+1)	120	0.1	3

Table 6.2 Experimental results of micro-ED milling process.

S. no	Voltage (V)	Capacitance (µF)	Feed rate (µm/s)	MRR (mm³/ min)	TWR (mm³/ min)	SR (µm)
1	−1	−1	−1	0.00268	0.000262	0.238
2	1	−1	−1	0.0036	0.000733	0.215
3	−1	1	−1	0.00317	0.000123	0.757
4	1	1	−1	0.00306	0.000154	0.762
5	−1	−1	1	0.00215	0.000264	0.125
6	1	−1	1	0.0029	0.000139	0.256
7	−1	1	1	0.00524	0.000627	0.501
8	1	1	1	0.00477	0.000257	0.636
9	−1.681	0	0	0.0042	0.000086	0.372
10	1.681	0	0	0.00411	0.000122	0.516
11	0	−1.681	0	0.00211	0.000513	0.197
12	0	1.681	0	0.00399	0.000457	0.749
13	0	0	−1.681	0.00288	0.000098	0.500
14	0	0	1.681	0.00425	0.000122	0.429
15	0	0	0	0.00415	0.000123	0.520
16	0	0	0	0.00421	0.000128	0.464
17	0	0	0	0.00454	0.000104	0.475
18	0	0	0	0.00426	0.000107	0.473
19	0	0	0	0.00412	0.000122	0.448
20	0	0	0	0.00424	0.000106	0.455

$$MRR = \frac{\text{Volume of material removal}}{\text{Machining time}} \qquad (6.1)$$

$$TWR = \frac{\text{Eroded volume of electorde (tool)}}{\text{Machining time}} \qquad (6.2)$$

6.3 Results and discussion

6.3.1 Micro-ED milling process statistical analysis

Statistical analysis is used to analyze machining performance based on control factor combinations, estimate coefficients, assess model appropriateness, and assess the prediction of outcomes (Sivaprakasam et al., n.d.) Design-Expert software was used to create the central composite design with three factors and three responses (MRR, TWR, and SR). Table 6.2 shows the experimental findings of titanium alloy micro-ED milling. The coefficients of determination (R^2) for MRR, TWR, and SR are 0.9747, 0.9963, and 0.9577, respectively. The backward elimination process eliminates insignificant components. The ANOVA results for MRR, TWR, and SR are shown in Tables 6.3–6.5, respectively.

6.3.2 Mathematical models equations for MRR, TWR, and SR

In Eqs. (6.3)–(6.5), quadratic equations are employed to express the MRR, TWR, and SR of the micro-ED milling process. These established model equations are applied to estimate responses with a coded unit in the following way:

Model equation of MRR

$$Y_{MRR} = 4.210 \times 10^{-3} + 6.873 \times 10^{-5} \times A + 5.910 \times 10^{-4} \\ \times B + 3.554 \times 10^{-4} \times C - 2.813 \times 10^{-4} \times AB + 6.262 \\ \times 10^{-4} \times BC - 4.417 \times B^2 - 2.596 \times C^2 \tag{6.3}$$

Model equation of TWR

$$Y_{TWR} = 1.165 \times 10^{-4} + 2.215 \times 10^{-6} \times A - 1.071 \times 10^{-5} \\ \times B + 9.817 \times 10^{-6} \times C - 7.757 \times 10^{-5} \times AB \\ - 1.0578 \times 10^{-4} \times AC + 1.452 \times 10^{-4} \times BC + 1.702 \\ \times 10^{-4} \times B^2 \tag{6.4}$$

Model equation of SR

$$Y_{SR} = 0.45 + 0.036 \times A + 0.20 \times B - 0.042 \times C + 0.036 \times AC \\ - 0.0395 \times BC \tag{6.5}$$

According to the quadratic equation, capacitance (B), feed rate (C), and the interaction between A and B, B and C have an influence on MRR and voltage has little effects on MRR. TWR is influenced by main factors, the interaction of A and B, and A and C. Voltage A, capacitance B, feed rate C, and the interaction between A and B, B and C have effects on SR.

Table 6.3 ANOVA for MRR micro-ED milling.

Source	SS	DoF	MS	F-value	p-value Prob > F	
Model	1.3878×10^{-5}	7	1.98×10^{-6}	66.688	<0.0001	Significant
A: Voltage	6.451×10^{-8}	1	6.451×10^{-8}	2.169	0.1665	
B: Capacitance	4.770×10^{-6}	1	4.770×10^{-6}	160.471	<0.0001	
C: Feed rate	1.725×10^{-6}	1	1.725×10^{-6}	58.03	<0.0001	
AB	6.328×10^{-7}	1	6.328×10^{-7}	21.288	0.0006	
BC	3.137×10^{-6}	1	3.137×10^{-6}	105.53	<0.0001	
B^2	2.839×10^{-6}	1	2.839×10^{-6}	95.49	<0.0001	
C^2	9.807×10^{-7}	1	9.807×10^{-7}	32.98	<0.0001	
Residual	3.567×10^{-7}	12	8.86×10^{-7}			
Lack of fit	2.440×10^{-7}	7	3.49×10^{-8}	1.546	0.3263	Not significant
Pure error	1.127×10^{-7}	5	2.25×10^{-8}			
Cor total	1.4239×10^{-5}	19				

SS, sum of square; *DoF*, degree of freedom; *MS*, mean square.

Table 6.4 ANOVA for TWR micro-ED milling.

Source	SS	DoF	MS	F-value	p-value Prob > F	
Model	7.32×10^{-7}	7	1.045×10^{-7}	461.73	<0.0001	Significant
A: Voltage	6.697×10^{-11}	1	6.697×10^{-11}	0.2957	0.5965	
B: Capacitance	1.566×10^{-9}	1	1.566×10^{-9}	6.914	0.0220	
C: Feed rate	1.316×10^{-9}	1	1.316×10^{-9}	5.810	0.0329	
AB	4.594×10^{-8}	1	4.594×10^{-8}	202.88	<0.0001	
AC	8.94×10^{-8}	1	8.94×10^{-8}	394.74	<0.0001	
BC	1.686×10^{-7}	1	1.686×10^{-7}	744.84	<0.0001	
B^2	4.25×10^{-7}	1	4.25×10^{-7}	1876	<0.0001	
Residual	2.718×10^{-9}	12	2.264×10^{-10}			
Lack of fit	2.169×10^{-9}	7	3.099×10^{-10}	2.828	0.1351	Not significant
Pure error	5.48×10^{-10}	5	1.096×10^{-10}			
Cor total	7.347×10^{-7}	19				

Table 6.5 ANOVA for SR micro-ED milling.

Source	SS	DoF	MS	F-value	p-value Prob > F	
Model	0.6175	5	0.1235	63.44	<0.0001	Significant
A: Voltage	0.0176	1	0.0176	9.074	0.0093	
B: Capacitance	0.5534	1	0.5534	284.33	<0.0001	
C: Feed rate	0.0241	1	0.0241	12.410	0.0034	
AB	0.0101	1	0.0101	5.2157	0.0385	
BC	0.0120	1	0.0120	6.210	0.0259	
Residual	0.0272	14	0.0019			
Lack of fit	0.0240	9	0.0026	4.115	0.0668	Not significant
Pure error	0.0032	5	0.0006			
Cor total	0.6448	19				

6.3.3 Model adequacy testing

The micro-ED milling process of the generated response surface model was evaluated using residuals analysis by model adequacy testing. The ANOVA results show that the model for MRR, TWR, and SR of the micro-ED milling process is significant. The MRR residual plot shows uniformly distributed residuals, mostly straight lines, and no outliers (Fig. 6.2). The errors in Fig. 6.3 are dispersed randomly, implying that they are unrelated. The absence of any anomalous structure in Fig. 6.4 demonstrates that the model is true. This demonstrates that the model fits the experimental values well. The similar tendencies have been observed in the TWR and SR models.

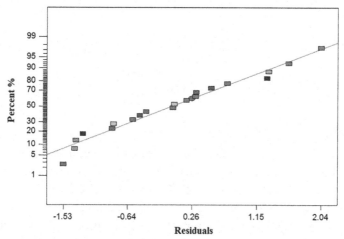

Fig. 6.2 Normal plot of residuals-MRR.

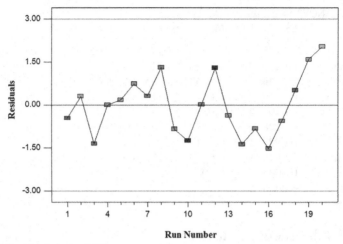

Fig. 6.3 Residual vs. run-MRR.

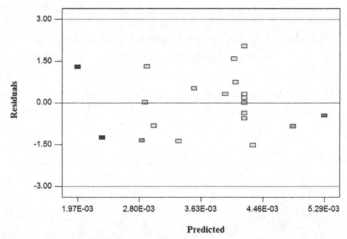

Fig. 6.4 Residuals vs. predicted values-MRR.

6.3.4 Analysis of MRR

The ANOVA was performed for analysis of input parameters on the MRR. The confidence levels model at 95%, which refers to p-values below 0.05. The model's R^2 score is 0.9747. The constructed model is satisfactory, and the estimated values match the experimental data well. The backward elimination process is used to eliminate the insignificant factors. Table 6.3 shows the reduced quadratic model for MRR of micro-ED milling.

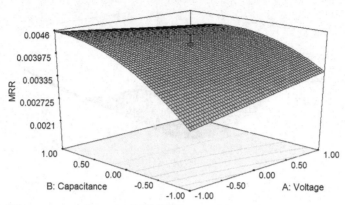

Fig. 6.5 3D response surface plot-MRR (AB).

Figs. 6.5 and 6.6 indicate that when capacitance and voltage increase, the MRR increases. The MRR depends on the discharge energy of process $E = 0.5 \ CV^2$. At high level of voltage and capacitance, the MRR has tendency to decrease. This is due to an increase in the spark gap at higher voltage and unwanted sparking and arching. The higher MRR was found at high levels of capacitance and feed rates. In addition, as seen in Fig. 6.6, the MRR tends to increase as feed rates are increased. The capacitance has the most influencing factors for micro-ED milling process when compared to other factors.

6.3.5 Analysis of TWR

ANOVA was performed for the analysis of input parameters on the TWR as shown in Table 6.4. The confidence levels model at 95%, whose probability

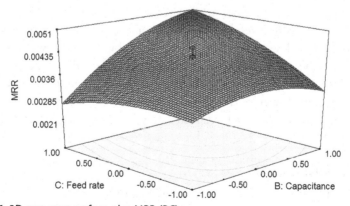

Fig. 6.6 3D response surface plot-MRR (BC)

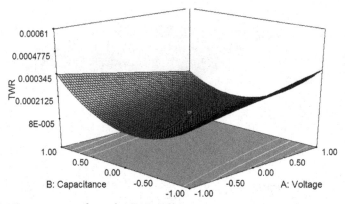

Fig. 6.7 3D response surface plot-TWR (AB).

value was lower than 0.05. The R^2 value of model is 0.9963. Fig. 6.7 shows that the excessive tool wear has found at low voltage and high capacitance and high voltage and lower capacitance, whereas the minimum tool wear has found at medium level of capacitance while keeping the feed rate at an average value. From the experimental results, it has found that there are very small variations in tool wear rate at all levels of voltage. The voltage controls the spacing between the electrodes, but once it reaches a certain point, it has no effect on TWR. There is a slight incremental trend while increasing the feed rate from lower to higher levels. The capacitance has a high impact on TWR; it was found that the minimum TWR was $0.000102\,\mathrm{mm^3/min}$ at the medium level of capacitance $(0.01\,\mu\mathrm{F})$, whereas at low and high level of capacitance, the TWR was $0.000297\,\mathrm{mm^3/min}$ and $0.000275\,\mathrm{mm^3/min}$, respectively.

Fig. 6.7 shows the 3D plot of TWR [AB]; at lower level capacitance $(0.0001\,\mu\mathrm{F})$, TWR increases while increasing voltage; at higher level of capacitance $(0.1\,\mu\mathrm{F})$, it tends to decrease while increasing the voltage from low level to high levels. Similar trends have exhibited at the lower-level feed rate $1\,\mu\mathrm{m/s}$; the TWR tends to increase and decrease at the higher level feed rate of $3\,\mu\mathrm{m/s}$, while increasing the voltage shown in Fig. 6.8, and Fig. 6.9 shows that at the lower level feed rate of $1\,\mu\mathrm{m/s}$, the TWR tends to decrease and increase at the higher level feed rate of $3\,\mu\mathrm{m/s}$, while increasing the capacitance.

6.3.6 Analysis of SR

Table 6.5 presents the findings of an ANOVA for SR with a 95% confidence level. The P-value shows the chance of a factor's significance. Figs. 6.10 and 6.11 show that when increasing voltage and capacitance, the SR values tend

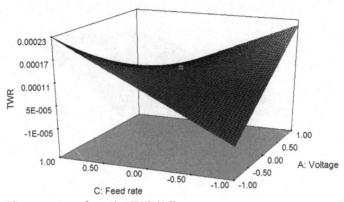

Fig. 6.8 3D response surface plot-TWR (AC).

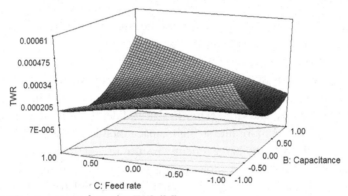

Fig. 6.9 3D response surface plot-TWR (BC).

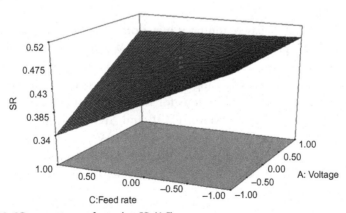

Fig. 6.10 3D response surface plot-SR (AC).

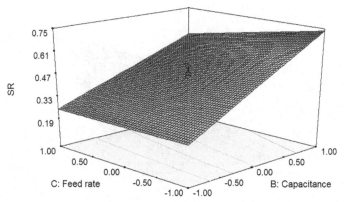

Fig. 6.11 3D response surface plot-SR (BC).

to increase. When increasing feed rate, the SR values were reduced. The capacitance is found to be the most influential factor in SR. The SR depends on the discharge energy of process $E = 0.5 \ CV^2$.

The energy deposited is determined by the capacitance value; thus, as capacitance increases, high discharge energy makes the more molten materials on the surfaces, and large crater in size and depth resulting in poor surface morphology. As a result, at high capacitance levels, the SR increases significantly. Lower discharge energy results in less intense energy and energy density, resulting in molten surface materials with tiny craters and peaks making smoother surface finish. Figs. 6.12 and 6.13 illustrate 2D roughness parameters and 3D surface image with medium- and high-level capacitance. Fig. 6.12 displays a relatively smooth surface roughness (Ra = 0.125 μm), whereas Fig. 6.13 exhibits a relatively rough surface (Ra = 0.762 μm).

The desirability-based multioptimization successfully employed for micro-WEDM process for better MRR, KW, and SR (Sivaprakasam, Hariharan, & Gowri, 2014, 2019). The optimal combination for micro-ED milling was achieved using a desirability-based multioptimization technique. The objective functions for micro-ED milling are the maximum MRR, and the minimum TWR and SR. Table 6.6 shows the micro-ED milling optimal conditions with 110V, 0.01 μF, and 3 μ/s. A confirmation test is used to validate the test results based on the optimal settings. The difference between the experimental and predicted values was less than 5%.

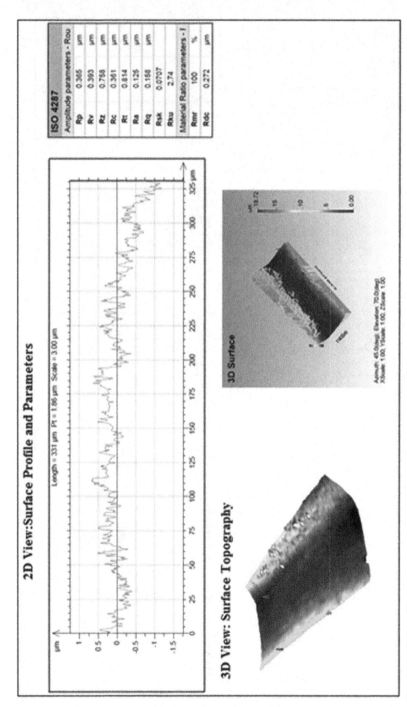

Fig. 6.12 2D Roughness parameters and 3D surface image (0.01 μF).

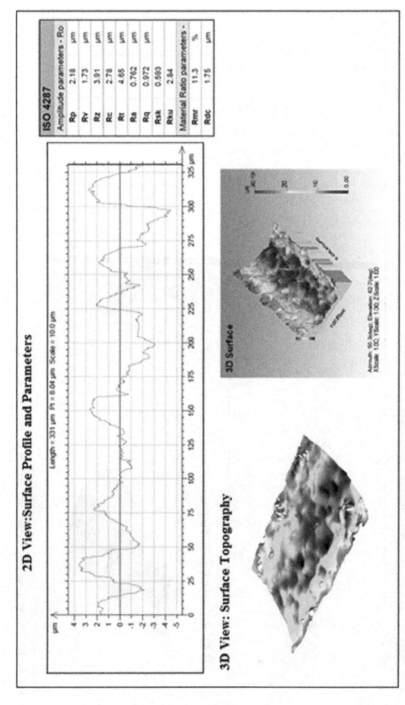

Fig. 6.13 2D Roughness parameters and 3D surface image (0.1 µF).

Table 6.6 Micro-ED milling optimal conditions.

Responses	Target	Desirability (D) value	A (V)	B (µF)	C (µ/s)	Predicted value	Experimental value
MRR	Maximum	0.814	110	0.01	3	0.00430	0.00426
TWR	Minimum	0.814	110	0.01	3	0.000126	0.000123
SR	Minimum	0.814	110	0.01	3	0.412	0.429

6.4 Conclusions

In this work, micro-ED milling has been attempted for the enhancement of machining performance on titanium alloy. The influences on the input factors were analyzed on MRR, TWR, and SR.

- Capacitance, feed rate, interaction of AB, BC, pure quadratic effect of capacitance (B^2), and feed rate (C^2) have more significant influence on MRR. However, voltage has less significant on MRR.
- The capacitance and feed rate have a high impact on TWR; it was found the minimum TWR of $0.000102 \, mm^3/min$ at the medium level of capacitance (0.01 µF). TWR was relatively unaffected by voltage. Interactions between voltage and capacitance, voltage and feed rate, and capacitance and feed rate have an effect on TWR. Pure quadratic effect of capacitance (B^2) has a significant influence on TWR.
- All three main factors and interaction of AB with BC have significant effects on SR.
- The higher MRR $(0.00426 \, mm^3/min)$ with lower TWR $(0.000123 \, mm^3/min)$ and SR $(0.429 \, µm)$ values was obtained using optimal machining conditions.

References

Ali, M. Y., Rahman, M. A., Zuhaida Zunairi, S. N., & Banu, A. (2017). Dimensional accuracy of micro-electro discharge milling. In *Vol. 184. IOP conference series: Materials science and engineering* Institute of Physics Publishing. https://doi.org/10.1088/1757-899X/184/1/012034. Issue 1.

Arun Pillai, K. V., & Hariharan, P. (2021). Experimental investigation on surface and machining characteristics of micro ED milling of Ti-6Al-4 V with different Nano powder mixed dielectrics. *Silicon, 13*, 1827–1837.

Bissacco, G., Hansen, H. N., Tristo, G., & Valentincic, J. (2011). Feasibility of wear compensation in micro EDM milling based on discharge counting and discharge population characterization. *CIRP Annals - Manufacturing Technology, 60*(1), 231–234. https://doi.org/10.1016/j.cirp.2011.03.064.

Bleys, P., Kruth, J. P., & Lauwers, B. (2004). Sensing and compensation of tool wear in milling EDM. *Journal of Materials Processing Technology, 149*(1–3), 139–146. https://doi.org/10.1016/j.jmatprotec.2003.11.042.

Cheng, X., Wang, Z., Kobayashi, S., Nakamoto, K., & Yamazaki, K. (2009). Development of a computer assistant programming system for micro/nano milling tool fabrication by multi-axis wire EDM. *International Journal of Computer Integrated Manufacturing, 22*(9), 847–856. https://doi.org/10.1080/09511920902866096.

D'Urso, G., Giardini, C., Lorenzi, S., Quarto, M., Sciti, D., & Silvestroni, L. (2020). Micro-EDM milling of zirconium carbide ceramics. *Precision Engineering, 65*, 156–163. https://doi.org/10.1016/j.precisioneng.2020.06.002.

Feng, W., Chu, X., Hong, Y., & Zhang, L. (2018). Studies on the surface of high-performance alloys machined by micro-EDM. *Materials and Manufacturing Processes, 33*(6), 616–625. https://doi.org/10.1080/10426914.2017.1364758.

Huang, H., Bai, J. C., Lu, Z. S., & Guo, Y. F. (2009). Electrode wear prediction in milling electrical discharge machining based on radial basis function neural network. *Journal of Shanghai Jiaotong University (Science), 14*(6), 736–741. https://doi.org/10.1007/s12204-009-0736-5.

Jafferson, J. M., Hariharan, P., & Ram Kumar, J. (2014). Effects of ultrasonic vibration and magnetic field in micro-EDM milling of nonmagnetic material. *Materials and Manufacturing Processes, 29*(3), 357–363. https://doi.org/10.1080/10426914.2013.872268.

Karthikeyan, G., Garg, A. K., Ramkumar, J., & Dhamodaran, S. (2012). A microscopic investigation of machining behavior in μeD-milling process. *Journal of Manufacturing Processes, 14*(3), 297–306. https://doi.org/10.1016/j.jmapro.2012.01.003.

Karthikeyan, G., Ramkumar, J., Dhamodaran, S., & Aravindan, S. (2010). Micro electric discharge milling process performance: An experimental investigation. *International Journal of Machine Tools and Manufacture, 50*(8), 718–727. https://doi.org/10.1016/j.ijmachtools.2010.04.007.

Koyano, T., Sugata, Y., Hosokawa, A., & Furumoto, T. (2017). Micro electrical discharge machining using high electric resistance electrodes. *Precision Engineering, 47*, 480–486. https://doi.org/10.1016/j.precisioneng.2016.10.003.

Kuriachen, B., & Mathew, J. (2016). Effect of powder mixed dielectric on material removal and surface modification in microelectric discharge machining of Ti-6Al-4V. *Materials and Manufacturing Processes, 31*(4), 439–446. https://doi.org/10.1080/10426914.2015.1004705.

Meena, V. K., & Azad, M. S. (2012). Grey relational analysis of micro-EDM machining of Ti-6Al-4V alloy. *Materials and Manufacturing Processes, 27*(9), 973–977. https://doi.org/10.1080/10426914.2011.610080.

Moses, M. D., & Jahan, M. P. (2015). Micro-EDM machinability of difficult-to-cut Ti-6Al-4V against soft brass. *International Journal of Advanced Manufacturing Technology, 81*(5–8), 1345–1361. https://doi.org/10.1007/s00170-015-7306-9.

Nakagawa, T., Yuzawa, T., Sampei, M., & Hirata, A. (2017). Improvement in machining speed with working gap control in EDM milling. *Precision Engineering, 47*, 303–310. https://doi.org/10.1016/j.precisioneng.2016.09.004.

Narasimhan, J., Yu, Z., & Rajurkar, K. P. (2005). Tool wear compensation and path generation in micro and macro EDM. *Journal of Manufacturing Processes, 7*(1), 75–82. https://doi.org/10.1016/S1526-6125(05)70084-0.

Pham, D. T., Dimov, S. S., Bigot, S., Ivanov, A., & Popov, K. (2004). Micro-EDM—Recent developments and research issues. *Journal of Materials Processing Technology, 149*(1–3), 50–57. https://doi.org/10.1016/j.jmatprotec.2004.02.008.

Pramanik, A., Basak, A. K., & Prakash, C. (2019). Understanding the wire electrical discharge machining of Ti6Al4V alloy. *Heliyon, 5*(4). https://doi.org/10.1016/j.heliyon.2019.e01473, e01473.

Qin, Y. (2015). *Micro-manufacturing engineering and technology.* Elsevier Inc.

P. Sivaprakasam, P. Hariharan, G. Elias. (n.d.). Experimental investigations on magnetic field-assisted micro-electric discharge machining of inconel alloy. International Journal of Ambient Energy, In press.

Rajurkar, K. P., & Lin, G. (2011). Review on electrical discharge machining of titanium alloy. In *International Conference on Precision, Meso, Micro and Nano Engineering (COPEN-7:2011)*. Pune: College of Engineering.

Sivaprakasam, P., Hariharan, P., & Gowri, S. (2014). Modeling and analysis of micro-WEDM process of titanium alloy (Ti–6Al–4V) using response surface approach. *Engineering Science and Technology, An International Journal, 17*(4), 227–235. https://doi.org/10.1016/j.jestch.2014.06.004.

Sivaprakasam, P., Hariharan, P., & Gowri, S. (2019). Experimental investigations on nano powder mixed Micro-wire EDM process of inconel-718 alloy. *Measurement, 147*, 106844. https://doi.org/10.1016/j.measurement.2019.07.072.

Yeo, S. H., Aligiri, E., Tan, P. C., & Zarepour, H. (2009). A new pulse discriminating system for Micro-EDM. *Materials and Manufacturing Processes, 24*(12), 1297–1305. https://doi.org/10.1080/10426910903130164.

Yu, Z. Y., Masuzawa, T., & Fujino, M. (1998). Micro-EDM for three-dimensional cavities—Development of uniform wear method. *CIRP Annals, 47*(1), 169–172. https://doi.org/10.1016/s0007-8506(07)62810-8.

CHAPTER SEVEN

Computational analysis of provisional study on white layer properties by EDM vs. WEDM of aluminum metal matrix composites

M. Kathiresan[a], R. Theerkka Tharisanan[b], and P. Pandiarajan[b]
[a]Department of Mechanical Engineering, E.G.S. Pillay Engineering College, Nagapattinam, India
[b]Department of Mechanical Engineering, Theni Kammavar Sangam College of Technology, Theni, India

7.1 Introduction

Metal matrix composites (MMC) are widely used in many applications because of their admirable properties such as high strength, high stiffness, high hardness, high damping capacity, good wear and corrosion resistance, and high thermal and electrical conductivities. MMC are manufactured with a combination of metals, organic compounds, or ceramics. Proper combination of two distinct material components performs better properties. The most preferable matrixes are aluminum, magnesium, titanium, nickel, and chromium, and reinforcements are used as ceramic particles like boron carbide (B_4C), silicon carbide (SiC), alumina (Al_2O_3), graphite etc. (Casati & Vedani, 2014; Kumar & Venkatesh, 2019).

Among the various matrix materials, aluminum alloys are majorly preferred for the production of MMC due to their unique properties. These properties comprise low density, high strength, good corrosion and wear resistance, good thermal conductivity, fatigue resistance, comparably low cost of manufacturing, and ease to be rolled, shaped, extruded, and drawn for desired postures. The low-strength and low-ductility aluminum alloys strengthened by the addition of reinforcements are dispersed in the matrix materials. The inclusion of reinforcements in the matrix aims to upgrade the properties of aluminum metal matrix composites (AMMC) macroscopic and microscopic levels (Ramnath et al., 2014; Sharma, Jhap, Kakkar, Kamboj, & Sharma, 2017). The strengthened AMMC find boundless applications, which include automobile, aerospace, marine, electrical and electronics, mining, sports, medical, and high-temperature applications (Orhadahwe, Ajide, Adeleke, & Ikubanni, 2020; Rohatgi, 1991).

Based upon types and particles size of reinforcements, working temperatures, and morphological conditions, AMMC are fabricated by various techniques such as squeeze casting, stir casting, powder metallurgy, liquid infiltration, and spray deposition. Among the various manufacturing processes, stir casting is mostly desirable because of low-cost manufacturing, achievements of homogeneous mixtures, working in high temperatures, and attempt to various selection processes of materials and manufacturing conditions (Kalaiselvan, Murugan, & Parameswaran, 2011; Pravin & Kavin Raj, 2019; Ramanathan, Krishnan, & Muraliraja, 2019).

Machining of AMMC is a serious factor because of high-loading and high-temperature applications environment. As a result of the presence of hard ceramic particles in the aluminum metal matrix composites, conventional machining is not suitable for fabricating the composites.

Hence, non-conventional machining process is mostly adopted to AMMC. Among the various non-conventional machining processes, electrical discharge machining (EDM) is largely preferred owing to its accurate machinability and versatile ability to machine the complicated and intricate shapes. The EDM is originated by the repetitive electric sparks between workpiece and electrode within the dielectric submerged medium. EDM is not making the contact between workpiece and electrode where it can dispose of vibration problems, chatter, and mechanical stresses. So that the EDM process is preferred for mold and die-making process and finishing parts for automobile, aerospace, and surgical components (Mohd Abbas, Solomon, & Fuad Bahari, 2007; Pramanik et al., 2020; Sidhu, Batish, & Kumar, 2013).

Due to the spark erosion process, the controlling gap between workpiece and electrode and discharge energy of plasma arc play a considerable effect on surface nature of the machined surface. From the literature, the maximum discharge energy associated with applied current, and controlling gap between workpiece and tool are the various parameters on EDM machining of EDM to obtain a smooth surface (Lin, Chen, Lin, & Tzeng, 2008; Liu, Huang, & Li, 2011).

Wire cut electrical discharge machining (WEDM) is involved to make the complex shapes and intricate profiles of applications such as aerospace, automobile, medical industries, and sports sectors. The material removal takes place by the thin wire through the melting and evaporation of workpiece. The wire traverse is monitored by the numerical control system to achieve the desired net shape accuracy. Many researchers have done a detailed investigation on the WEDM control parameters such as voltage, current, pulse duration, wire diameter, and wire material. Machined surface of WEDM mainly depends upon the discharge of arc through the controlled environment (Bobbili, Madhu, & Gogia, 2015; Devarasiddappa & Chandrasekaran, 2020; Khan, Khan, Siddiquee, & Chanda, 2014; Lau & Lee, 1991). It has been remarkable that most of the work is engaged with surface modification of EDM workpiece with various machining parameters and none of the work is conducted on surface modification of the machined workpiece compared with EDM and WEDM.

7.2 Heat-affected zone (HAZ) on EDM

EDM takes a part of thermal erosion process between the workpiece and the electrode through diverse electrical sparks. In this situation, the workpiece and the electrode are immersed in dielectric fluid. The thermal energy develops high-energy plasma between the workpiece and the electrode at a temperature from 8000°C to 20,000°C. This consequential amount of heat helps to melt the material at the surface of the product (Ho & Newman, 2003). Enduring improvements in EDM has been achieved by continuous MRR and numerical control process parameters. Crater formation, microcracks, resolidified metal droplet, chimneys, and pockmarks are certain issues in EDM process. The uncontrolled electric discharge causes the material and also evaporates the dielectric fluid. Only 15% or less amount of molten material is carried away by dielectric and the balance metal is deposited on new machined surface as undulating terrain (Zhang, 2014). This problem is seriously considered to enhance the properties of AMMC during EDM process. EDM process directly affects the various layers in the surface of the workpiece, which is referred as altered metal zone (AMZ) or heat-affected zone (HAZ). Fig. 7.1 represents the various layers such as white layer and HAZ on machined surface. An aerospace, tool, and die industry is focused on metal removal and quality of surface finish. The components have to be cleaned and have good surface characteristics. Workpieces made of aluminum metal matrix composites suffer thermal damage caused by an incipient pulse discharge temperature gradient (Choudhary, Kumar, & Garg, 2010).

Various analytical models are developed for the estimation of MRR in EDM with consideration of thermal energy as medium. Unfortunately, the

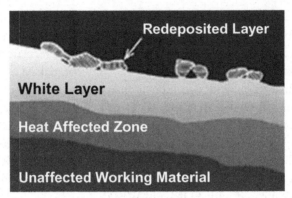

Fig. 7.1 White layer thickness and heat-affected zone.

prediction of MRR is enormously varied with the actual one. Due to the remarkable process, parameters such as pulse-on time, pulse-off time, current, voltage, dielectric fluid, types of electrode, and the gap between workpiece and electrode are involved to affect the machined surface and make the white layer on machined surface. This layer formed about 2.5–50 μm as sticky part on the machined surface. It is excessively porous, microcracks, hard, and brittle. The affected machined surface by re-cast layers causes the EDM application product as incompatible and poor. This work is engaged with analysis of the HAZ and subsurface analysis (SSA) like white layer thickness (WLT) through the fixed EDM machining parameters.

7.3 Heat-affected zone (HAZ) on WEDM

The cutting-edge research areas of automobile, aerospace, missiles, nuclear reactors, and ceramic industries are engaged with qualitative characteristics of surface parameters in high-temperature applications. The wire-cut electrical discharges machining procedure seeks to minimize residual thermal stresses in surface products while achieving the best quality judgment (Singh & Misra, 2016). The desired machining parameters are mostly involved to predict the WEDM responses like MRR and surface roughness. The quality assessment of MRR and SR is directly affected by HAZ on metal and wire electrode meeting point. On this wise, the HAZ is directly connected with improvement of MRR and SR by predicting the desirable machining parameters. Due to the electrical erosion process, pertaining to thermal process, the surface structure can be modified on HAZ. This

HAZ of WEDM and its performance characteristics are firmly connected to the lifetime and durability of the product (Straka, Čorný, & Piteľ, 2016).

7.4 White layer on EDM and WEDM

In EDM, the high temperatures of dense thermal sparks contribute to develop the clean surface through the erosion process. Due to these temperature changes in material, it causes metallurgical changes in workpiece. This secondary phase changes in EDM are involved to develop the recast layer on machined surface, which is called as "white layer." This white layer is the development of the solidification of a melted material, and it exhibits a brittle nature. Sometimes, it shows higher strength, higher hardness, better bonding with parent material, and also high corrosion resistance. Though, it accommodates the pore hole, small crack, and globules on subsurface of the material, which causes the lower life of the product and appears poor surface finish (Cusanelli et al., 2004).

The reality of white layer during EDM process has been known for 30 years, but the comparison study of EDM and WEDM on the white layer thickness development has not taken seriously. Various control factors of EDM such as current, voltage, clearance between the workpiece and tool, pulse-on time, pulse-off time, type of electrode, and type of dielectric fluid are used. Among these parameters, dielectric is important because spark gap is filled during machining process and also reacts with chemical process. The water-in-oil (W/O) emulsion dielectric creates a good surface and thin white layer thickness than deionized water and kerosene (Zhang, Liu, Ji, & Cai, 2011). In addition, the surface roughness is decided by erosion and pulse current during the machining process. Additionally, the desired polished surface finish is achieved by abrasive machining process.

Ahmet Hascalik and Ulas Coydas conducted an experiment with the EDM and electrochemical grinding with metal-bonded abrasive tool (AECG) on Ti-6Al-CW alloy. In this experiment, the surface roughness was affected by current, and also gentle surfaces were achieved from EDM process through the processing of AECG by concentrate, the electrolyte flow rate, machining voltage, and feed rate (Hasçalık & Çaydaş, 2007).

A WEDM is used to cut the electrically conducting material without touching the workpiece by generating the thermal sparks. During the erosion process, metallurgical has been occurred due to the plastic deformation under high temperature. This HAZ in WEDM takes the serious one because of microstructure of material and surface integrity was offered by unfitted machining parameters. The quality of the eroded surface in WEDM is comparable with EDM-machined surface through the analysis of HAZ and white layer parameters. The high quality of material is achieved by conductibility of heat to the parent material. The melted material surface carries small amount of oxygen, hydrogen, and few melted off burs from electrode. Experimental values of HAZ in machined surface can be determined through the microhardness measurements. The HAZ depth and surface layer formation are closely fallen to life of the product (BorZíkoVA, 2009).

7.5 Thermophysical model for EDM

The thermal damage occurring in the EDM process due to uncontrolled side sparks, tool electrode shape, degradation and poor workpiece cavity, and the surface that appeared in the workpiece is not a mirror image. EDM is also influenced to develop the unfavorable tensile residual stresses due to cause of the high temperature gradient by pulse discharge. Hence, the study of thermophysical model for EDM is essential one. The analytical model gives the precise result of MRR to the experimental data (Ming et al., 2014). This mechanism of the metal removal model helps to identify the HAZ during the machining process in EDM. This thermophysical two-dimensional model is applied between tool electrode and workpiece. The boundary conditions and heat source model are shown in Fig. 7.2.

The discharge of the EDM can be originated by the proper plasma channel. The following governing Eq. (7.1) is used to find the transfer of heat source from the electrode to the workpiece ($Z = 0$). Here, the disk-shaped heat source model is considered (Liu & Guo, 2016; Pandey & Jilani, 1986).

$$k \left[\frac{1}{r} \frac{\partial}{\partial r} \left(r \frac{\partial T}{\partial r} \right) + \frac{\partial^2 T}{\partial z^2} \right] = \rho C \frac{\partial T}{\partial t} \tag{7.1}$$

The surface region narrated by $Z = 0$ and the constant heat flux have been assumed by taking of $0 \leq r \leq R$, whereas the heated surface has been considered to be excellently insulated. The boundary conditions for current disk-shape heat source models are shown in Eq. (7.2):

$$-K \frac{\partial T}{\partial z}(r, 0, t) = \begin{cases} Q \, for \, 0 \leq r \leq R \\ 0 \, for \, R \leq r \leq b \end{cases} \tag{7.2}$$

This equation is a simplified form of the analytical model to predict the MRR and surface roughness by concentrating on the heat source model.

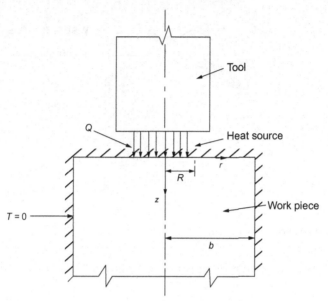

Fig. 7.2 Boundary conditions of heat source model for disk shape.

Furthermore, this model gives the near results to the experimental data. The author relates this analytical model to the heat-affected zone on EDM due to the assumption of the clear boundary conditions.

7.6 Thermophysical model for WEDM

The working mechanism of WEDM is still at large under research. Various analytical modeling methods are used to predict the working mechanism of WEDM machining process. The energy transfer between the tool electrode and the workpiece is controlled by the control factors such as current supplied and pulse duration. This numerical model is considered to study the total discharge energy transfer between the anode and the cathode medium. Several researchers contributed to create and check the numerical model with experimental results (Mohapatra, Sahoo, & Bhaumik, 2016; Singh, Das, & Sarma, 2018). The purpose of this work is to investigate the implications of thermal damage on WEDM process through the use of thermomechanical prediction.

This prediction illustrated 3D model involved to understand the accurate results on HAZ and their sublayers by applying the proper boundary conditions to the model as shown in Fig. 7.3. The heat transfer can be occurred

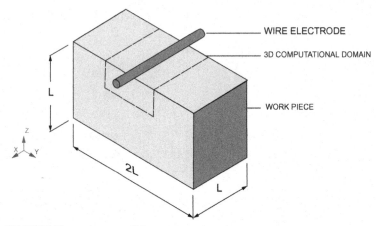

Fig. 7.3 WEDM boundary conditions.

in xy plane, and assume there is no heat transfer obtained in z direction, which is described in Fig. 7.4.

The above schematic, HAZ, is far away from the left, right, and bottom sides of the boundary conditions. The small plasma channel will be delivered in top side of the boundary conditions and other sides of the boundary conditions are considered as constant temperature. The severe plasma is involved to fix the decent metal removal by heating and erosion of particles.

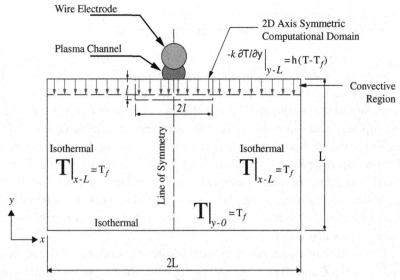

Fig. 7.4 WEDM boundary conditions for heat transfer.

Due to the uncontrolled heat conduction, the surface damage occurred. The prediction heat condition to the workpiece considered by the thermophysical model is represented in Eq. (7.3).

$$\rho \, c_p \, \frac{\partial T}{\partial t} = \frac{\partial}{\partial x}\left(k \, \frac{\partial T}{\partial x}\right) + \frac{\partial}{\partial y}\left(k \, \frac{\partial T}{\partial y}\right) \tag{7.3}$$

where thermophysical properties like the density (ρ), specific heat (C_p), and thermal conductivity (k) are considered autonomous of temperature.

In the present work, the thermal damage of a WEDM process has been analyzed for high-temperature applications. This numerical model has been considered to find out the accurate results by taking the authentic boundary systems. By considering the HAZ and their sublayers, the AMMC properties can be retained through the optimized process temperature.

7.7 Methodology of composite making

7.7.1 Chemical elements of pure Al6061 perimental material

In order to investigate the white layer properties, the AMMC workpieces are used with two different unconventional machining such as EDM and WEDM. In this work, the aluminum alloy 6061 (Al6061) was used as matrix material, and the chemical composition of the matrix material is illustrated in Table 7.1. The presence of various major elements such as Mg and Si is confirmed by energy-dispersive X-ray spectroscopy (EDAX), as shown in Fig. 7.5.

The mechanical and machining properties are improved by the different types of reinforcements. Here, the fly ash (coal ash) is used as a reinforcement particle. The fly ash consists of fine powders and is predominantly spherical in shape. The chemical composition of fly ash particles is shown in Table 7.2.

7.7.2 Experimental procedure

Undoubtedly, the whole property of the AMMC depends upon how it was fabricated. Stir casting is one of the promising routes used in AMMC manufacturing as the mechanical properties of the prepared AMMC are

Table 7.1 Chemical composition of Al6061 alloy.

Element	Mg	Si	Fe	Cu	Ti	Cr	Zn	Mn	Al
%	0.90	0.75	0.25	0.22	0.09	0.10	0.05	0.04	Reminder

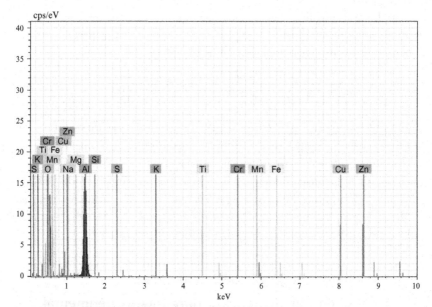

Fig. 7.5 EDAX analysis of pure Al6061.

Table 7.2 Chemical composition of fly ash.

Element	% of composition
SiO_2	50.2–59.7
Al_2O_3	14–32.4
Fe_2O_3	2.7–16.6
CaO	0.6–9
K_2O	0.2–4.7
MgO	0.1–2.3
SO_3	Na
TiO_2	0.3–2.7
Na_2O	0.2–1.2
P_2O_5	Na
MnO	Na
LOI	0.5–7.2

far better than the other manufacturing methods. The continuous stirring action helps to make the AMMC as decent homogeneous and less defective.

In this method, aluminum 6061 billet was cut and melted in graphite crucible furnace up to whole metals come to liquid phase. After that, the Al6061 metal was reached in 700°C, and the stirrer was introduced to make the homogeneous mixture before adding the reinforcements into the molten

Fig. 7.6 Stir casting set up.

matrix metal. The stir casting setup is illustrated in Fig. 7.6. Subsequently, the fly ash reinforcement powder was preheated around 250°C in preheater furnace. This preheat task has been to make the homogeneous mixture and high wettability. The stirrer speed was gradually decreased while the reinforcement powder was progressively added. In order to improve the wettability, the small amount of magnesium (Mg < 1 wt%) and also gas tablets like K2TiF6 has been added in the molten AMMC (Patil, Ansari, Selvan, & Thakur, 2021; Poovazhagan, Rajkumar, Saravanamuthukumar, Javed Syed Ibrahim, & Santhosh, 2015). In this process, the mechanical stirring action has been taken place to create the vertex at 800 rpm in maintaining the furnace temperature of 750°C for 10–15 min. Thereafter, the stirrer speed was slightly decreased by maintaining the constant temperature of 700°C for 5 min. In this stage, the stirrer speed was increased again to 800 rpm by maintaining the same temperature level for 10 min. Finally, the molten AMMC was poured into a preheated die for solidifying in environmental cooling. This procedure was followed for the entire samples such

Table 7.3 Weight proportions of AMMC.

Sample specifications	Al6061 wt%	Fly ash (coal ash) wt%
S1	Pure Al6061	–
S2	95	5
S3	90	10

as S1, S2, and S3. The sample proportions and descriptions are illustrated in Table 7.3.

7.8 Machining setup

The near net shape of the product is possible by combining different manufacturing methods. However, it is challenging to machine the hard AMMC due to the dimensional tolerance and component design. Basically, AMMC are hard to machine materials through the medium of hard ceramics reinforcements and their observation in nature. Non-conventional machining processes like EDM and WEDM are frequently used processes for achieving the required quality surface finish, low tool wear rate, and high MRR and bring about a complicated manufacturing shape (Chaitanya Reddy & Venkata Rao, 2020; Dey, Debnath, & Pandey, 2017; Suthar & Patel, 2018).

7.8.1 EDM machining

EDM machining, a (ELECTRONICA—ELPULS) machine, was used with copper electrode of 20 mm diameter. The EDM machining parameters are illustrated in Fig. 7.7.

The aim of EDM machining is to study the HAZ on machined surfaces and the sublayers such as white layer thickness and globules diameter at different setting control parameters of current (I), pulse-on time (TON), and pulse-off time (TOFF). EDM machining was performed on three samples, namely, Pure Al6061, Al6061 + 5 wt% FA, and Al6061 + 10 wt% FA, with constant machining depth fixed as 2 mm all over the experiments (Uthayakumar et al., 2019). The EDM workpiece was cut in to 100 mm × 100 mm × 10 mm thick for machining.

7.8.2 WEDM machining

WEDM is widely used in industrial applications due to machining of the MMC and cemented carbides. In this work, the (ELECTRONICA—

Fig. 7.7 EDM machine.

ELEKTRA—ELPLUS 40A DLX) machine was engaged with machining of AMMC by copper wire (0.1–0.3 mm) (Garg et al., 2010). The arrangements of WEDM machine are shown in Fig. 7.8. The three samples such as Pure Al6061 (S7), Al6061 + 5 wt% FA (S8), and Al6061 + 10 wt% FA (S9) were chosen as workpieces with the dimension of 100 mm × 100 mm × 10 mm. The WEDM specifications are illustrated in Table 7.4.

Table 7.4 WEDM machine specifications.

Experimental parameters	Specifications
EDM machine	ELECTRONICA—ELEKTRA –ELPLUS 40A DLX
Tool electrode	Copper wire electrode (0.1–0.3 mm diameter)
Dielectric fluid	Kerosene oil
Machining conditions	Current (ampere) = 10 pulse-on time (µs) = 100 pulse-off time (µs) = 17

Fig. 7.8 WEDM machine.

7.9 Results and discussion

7.9.1 White layer and globules on EDM

After EDM machining, the workpieces were cleaned and dried for measuring the HAZ, white layer thickness, and globules diameter using the SEM machine TESCAN VEGA 3. The excess amount of heat generated the undesirable sparks on EDM machining surface of pure Al6061 sample, which causes the damage on machined surface as shown in Fig. 7.9. Consistently in EDM process, the primary layer on the machined surface is called white layer (WL), and the secondary layer followed below the primary layer is called HAZ, whose thickness could be appeared between 5 and 200 μm (Thao & Joshi, 2008). The direction of this investigation is to identify

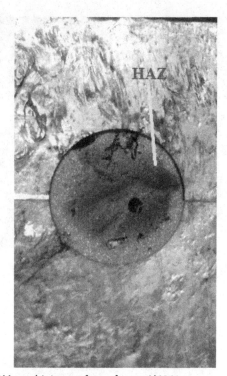

Fig. 7.9 HAZ on EDM machining surface of pure Al6061.

and study the effect of machining conditions on HAZ and WL of pure aluminum 6061 and AMMC. Generally, the EDM-machined surfaces were obtained with low glossy appearance wrapped by pock marks, globules of debris, shallow craters, and white layers constituted by entrapped gases escaping from the re-casted or re-solidified or re-deposited material (Lee, Lim, Narayanan, & Venkatesh, 1988). Fig. 7.10 shows the typical white layer taken from the pure Al6061 and AMMC machined through EDM. The SEM images are used to generate high-resolution images to identify the HAZ and recast layers and globules diameters (GD). The SEM images are contributed to investigate the appearance of the HAZ, WL, and GD at the operating voltage of 20 kV with various resolutions. Table 7.5 illustrates that the area and length of the WL for various EDM-machined samples. The WL was clearly shown in Fig. 7.10 and the WLT was noted almost in all machining operations at a distance of 10 to 20 μm from the top of the EDM-machined surface. The major WLT was appeared on Al6061 + 5 wt % FA (S5) samples due to the excess molten material deposited on the surface. Afterward, the HAZ was appeared on the machined surface formed as

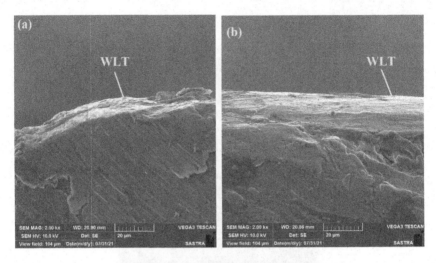

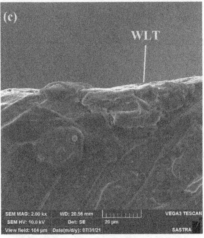

Fig. 7.10 WLT on EDM samples (A) Pure Al6061. WLT on EDM samples (B) Al6061 + 5 wt % FA. WLT on EDM samples (C) Al6061 + 10 wt% FA.

hardened layer during the EDM machining process. The copper tool was slightly melted on the machined surface. Fig. 7.10 exhibits an appearance of surface cracks and globules on the cross section of pure 6061 and two various proportions. The small cracks were appeared very large on Al6061 + 10 wt% FA compared with other two samples, and also it was extended in a perpendicular direction to the machined surface and continued to bulk metal of AMMC (Lin et al., 2008). The GD on EDM-machined samples are as follows in Fig. 7.11.

Table 7.5 Measurement of WLT on various EDM machined samples.

Sample number	Proportion	Area of WLT in μm^2	Length of WLT in μm
S4	Pure Al6061	1.444	10.612
		1.092	7.896
		1.425	10.344
		1.296	9.524
S5	Al6061 + 5 wt% FA	1.444	10.49
		1.388	10.12
		1.259	9.074
		1.166	8.376
S6	Al6061 + 10 wt% FA	0.904	6.552
		0.271	1.897
		0.289	2.069
		0.163	1.034

Typically, the fatigue life of the AMMC component is characterized by the following factors such as surface cracks, surface morphology, surface roughness, and residual stresses on machines surface. The repetitive sparks produce the decent surface roughness by dissipation of heat to the workpiece and dielectric fluid. Due to the poor solidification during the machining process, the globules are involved to create the crack initiation and hard brittle in nature (Arooj, Shah, Sadiq, Jaffery, & Khushnood, 2014). The GD for the various EDM-machined samples are listed in Table 7.6.

The small and large globules are appeared on almost all samples. The higher GD was obtained on pure Al6061 as 11.626 μm. The lower GD was generated on Al6061 + 10 wt% FA samples as 2.087 μm. This GD

Table 7.6 GD for various EDM machined samples.

Sample number	Proportion	Area of WLT in μm^2	Length of WLT in μm
S4	Pure Al6061	1.6	11.626
		0.276	1.849
		0.386	2.75
		0.313	2.179
S5	Al6061 + 5 wt% FA	0.179	1.882
		0.211	2.224
		0.39	4.277
		0.211	2.245
S6	Al6061 + 10 wt% FA	0.478	3.323
		0.386	2.717
		0.423	2.992
		0.294	2.087

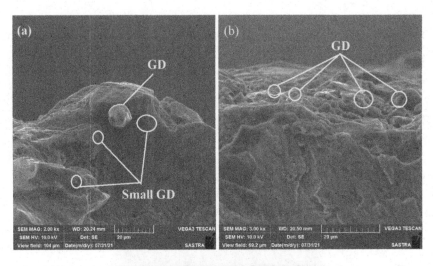

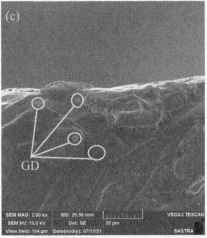

Fig. 7.11 GD of EDM-machined samples (A) Pure Al6061. GD of EDM-machined samples (B) Al6061 + 5 wt% FA. GD of EDM-machined samples (C) Al6061 + 10 wt% FA.

indicated that the poor dielectric flush and rapid solidification cause the machining parameters due to the temperature gradient on the EDM-machined surface. Furthermore, the string spark (or) strength of the spark involved to increase the more molten metal on the machined surface. The SEM images show that the globules are dispersed on some of the areas on the work samples, which is obtained by the strong sparks produced by the plasma.

7.9.2 White layer and globules on WEDM

The renovation in the process design and accurate control of wire EDM (WEDM) grants the usage of this manufacturing technology for a variety of demanding applications. In order to examine the HAZ and sublayers, pure Al6061 and AMMC samples were processed by WEDM. All machining work was carried out in ELECTRONICA—ELEKTRA—ELPLUS 40A DLX. The hard copper wire of 0.1–0.33 mm diameter, elongation of 45 MPa, tensile strength of 390 MPa, and thermal conductivity of 370 W/mK are used for machining of pure Al6061 and AMMC under specific conditions.

The WEDM control parameters such as cutting speed and current were controlled by WEDM tool monitor. The machined work samples were cut from casted samples as 100 mm × 100 mm × 10 mm, and the machined sample of pure Al6061 is illustrated in Fig. 7.12.

As a result of very high temperature 10,000–20,000°C that appeared during WEDM, the surface of the work material is completely melted at cutting zone and suddenly cooled by dielectric fluid, and that layer is also called recast layer (or) white layer (Mouralova et al., 2020). The HAZ, WL, and GD are directly involved to ensure the physical and mechanical properties of the product. In this work, the WEDM HAZ and sublayers were determined at cross section of the work samples by SEM images, which are described in Fig. 7.13.

The WLT was measured in four different places along the white layer-formed region, and the corresponding WLT for EDM-machined samples are described in Table 7.7.

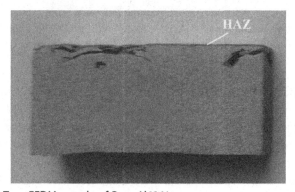

Fig. 7.12 WLT on EEDM sample of Pure Al6061.

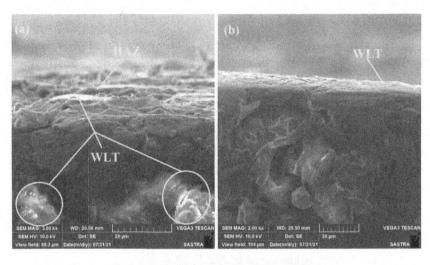

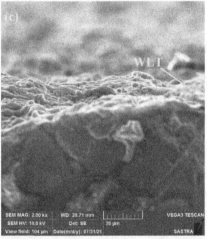

Fig. 7.13 WLT on WEDM samples (A) pure Al6061. WLT on WEDM samples (B) Al6061 +5 wt% FA. WLT on WEDM samples (C) Al6061 +10 wt% FA.

The surface roughness is confirmed on WEDM by producing the series of sparks in between the perfect location moving wire, and AMMC samples. Heavy frequency pulses are generated by alternating current (or) direct current, which decides the material to become smoother. And also, the following WEDM control parameters are observed by many researchers such as current, wire ratio, wire material, wire diameter, wire speed, kerf width, and dielectric fluid (Patel & Maniya, 2018). Owing to the fusion and vaporization process on WEDM, the AMMC was damaged. Therefore, the

Table 7.7 Measurement of WLT on various WEDM machined samples.

Sample number	Proportion	Area of WLT in μm^2	Length of WLT in μm
S7	Pure Al6061	0.747	8.276
		0.699	7.701
		0.627	6.897
		0.715	7.931
S8	Al6061 + 5 wt% FA	1.03	7.478
		0.956	6.957
		0.846	6.087
		0.772	5.565
S9	Al6061 + 10 wt% FA	2.758	20.174
		1.103	8
		0.791	5.739
		2.004	14.609

problem during the WEDM process is divided into three categories, namely, recast layer (or) white layer, globules diameter, and HAZ. The SEM microstructure investigation is contributed to determine the GD on various places of machined areas as shown in Fig. 7.14.

The globules characteristics need to be studied on various aspects of research works. The various constraints such as EDM control parameters, workpiece material, and electrode material should be accounted for determining the GD. The size of the GD for various EDM-machined samples is listed in Table 7.8.

The elevated GD was noted on Al6061 + 5 wt% FA as 20.493 μm and the least GD was determined on Al6061 + 10 wt% FA samples as 1.486 μm. The

Table 7.8 GD for various WEDM machined samples.

Sample number	Proportion	Area of WLT in μm^2	Length of WLT in μm
S7	Pure Al6061	1.168	12.907
		0.495	5.381
		0.454	4.964
		0.568	6.25
S8	Al6061 + 5 wt% FA	0.349	2.441
		0.736	5.298
		0.662	4.747
		2.795	20.493
S9	Al6061 + 10 wt% FA	0.367	6.775
		0.116	2.102
		0.116	2.074
		0.084	1.486

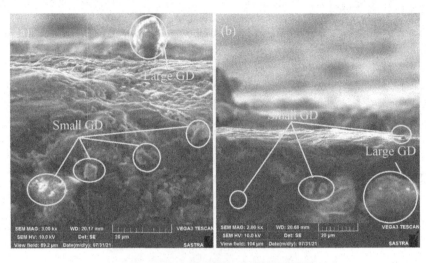

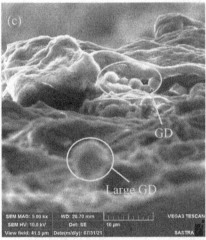

Fig. 7.14 GD of EDM-machined samples (A) pure Al6061. GD of EDM-machined samples (B) Al6061 + 5 wt% FA. GD of EDM-machined samples (C) Al6061 + 10 wt% FA.

average globule diameters for EDM and WEDM samples are shown in Figs. 7.15 and 7.16.

The AGD of EDM samples decreased from pure Al6061 to Al6061 + 5 wt% FA by the average value of 4.601 µm to 2.657 µm and after slightly increased to 2.779 µm on Al6061 + 10 wt% FA. Subsequently, the AGD of WEDM samples initially increased on pure Al6061 sample from 7.3755 µm to Al6061 + 5 wt% FA sample as 8.2448 µm and finally ended with 3.1093 µm on Al6061 + 10 wt% FA samples.

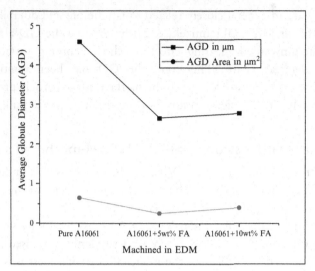

Fig. 7.15 Average globule diameters (AGD) for EDM samples.

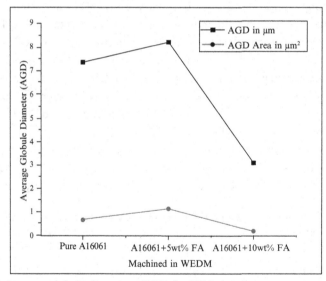

Fig. 7.16 Average globule diameters (AGD) for EDM samples.

The minimum number of researches has strived to explore the GD impact on the surface integrity of the samples. This GD mechanism is also unpredictable because of complex and debatable natures and more control parameters involved with the WEDM process. The overall area and length of the GD are in general closely related to fatigue life and durability of the product (Mouralova, Zahradnicek, & Hrdy, 2016). The GD also formed due to the unwanted shearing effect on the machining surface by the WEDM electrode wire. Therefore, the GD has been controlled on Al6061 + 10 wt% FA samples due to the hard nature of particle presentation in the AMMC, which significantly dissipate the heat to the bulk material.

7.9.3 Microhardness on EDM- and WEDM-machined samples

The microhardness or micro-Vickers hardness was conducted on KRYSTAL hardness machine. The microhardness of the various machined samples for EDM and WEDM is given in Table 7.9.

According to this, the micro-Vickers hardness on EDM samples was increased on pure Al6061 as 127 HV and suddenly decreased to 97 HV on Al6061 + 5 wt% FA. Finally, the micro-Vickers hardness was increased on Al6061 + 10 wt% FA samples as 107 HV. An increase of microhardness has been observed on AMMC in the heat affected zone. The minimum microhardness observed on WEDM samples as 89 HV on Al6061 + 10 wt % FA. Due to the high temperature of sparks and dielectric fluid, the property and composition of white layer are different than those of the parent material. This layer is divided into four parts for both EDM and WEDM after the machined parts. The first layer is called debris layer; the second layer is called white layer; the third layer is called annealed layer (or) HAZ; the fourth layer is called bulk material (or) parent material (Markopoulos, Papazoglou, & Karmiris-Obratański, 2020).

Due to the high temperature of sparks and dielectric fluid, the property and composition of white layer are different than those of the parent material. A small amount of ablated vaporized material is brittle in nature

Table 7.9 Micro-Vicker hardness (HV) on EDM and WEDM machined samples.

Machined samples	Pure Al6061	Al6061 + 5 wt % FA	Al6061 + 10 wt % FA
Micro-Vickers hardness (HV) on EDM	127	92	107
Machine	123	113	103
	119	109	108
	98	88	96
Micro-Vickers hardness (HV) on	134	143	89
WEDM Machine	129	141	98
	114	126	94
	109	101	87

as well as showing signs of creating cracks (or) breaking of the material by porosity and crater formation. In this manner, the thermal energy is transferred from WL to the parent material through the HAZ. Also, the presence of silicon dioxide (SiO2) is around 50 wt% on the coal fly ash, which acts as a thermal gradient barrier to the bulk material. The result concluded that the HV are not propogate beyond 21 μm from the top machined surface to the bulk material region in EDM and WEDM samples. The rapid cooling and optimized pulse energy can predict the possible phase changes and changes in the chemical composition in the AMMC (Kumar, Singh, Singh, & Sethi, 2009; Lin, Hwang, & Fung, 2016). Therefore, due to the heat flow energy transfer in HAZ, the microhardness is varied. The microhardness has been measured from WL to HAZ in between the small micrometer (μm) interval for EDM and WEDM, which are tabulated. This hard ductile nature of material forms the shielding effect on parent material through the protection from dissipated heat.

7.10 Conclusion

After an elaborate experimental study of the AMMC, the following conclusions are drawn.

- The Al6061–FA composites were produced successfully by stir-casting techniques with various reinforcement weight percentages (viz., 5 wt% and 10 wt%).
- The SEM and EDAX analyses revealed that the percentage of coal fly ash (FA) particles are evenly distributed in AMMC.
- The SEM microstructure confirmed the existence of the surface integrity such as recast layer (or) white layer, globules, and HAZ on EDM- and WEDM-machined samples.
- There is a WLT formed on the EDM- and WEDM-machined surfaces. The surfaces of WL are heavily rough and brittle nature due to the improper flashed debris during the machining process. Here, the control parameters for each machining condition followed the same value assurance.
- The SEM micrographs evidenced that the HAZ has been involved to develop the black marks, small pores, and small crack on the EDM-machined samples. Meanwhile, the work samples were damaged due to the pulse generation and applied current on EDM process.
- In EDM process, the large WLT was found on pure Al6061 sample as 10.612 μm and the short WLT was found on Al6061 + 10 wt%

FA sample as 1.034 μm. The higher GD was found on pure Al6061 as 11.626 μm.

- In WEDM process, the higher and lower WLT were determined on Al6061 + 5 wt% FA as 20.174 μm and 5.565 μm compared with other two sample proportions. Meanwhile, the higher globules were appeared on Al6061 + 5 wt% FA as 20.493 μm and the lower globules were obtained on Al6061 + 10 wt% FA as 1.486 μm.
- Finally, the microhardness test was conducted successfully. The higher microhardness was determined on pure Al6061 for each machined sample. Subsequently, the micro-Vickers hardness was gradually decreased from WLT region to bulk material region by crossing over the HAZ.
- The results concluded that the experimental study of HAZ, WLT, GD, and micro-Vickers hardness has been analyzed through the theoretical model for EDM and WEDM. The lifetime of the EDM- and WEDM-machined parts can be improved by selecting the optimized machining parameters such as pulse generation, pulse density, pulse strength, proper solidification during metal eroding stage, decent dielectric flushing, and applied current.
- The research works on EDM and WEDM using pure water as dielectric is a rising area for the sustainable development through green manufacturing process. AMMC are yet to be examined in these areas.

References

Arooj, S., Shah, M., Sadiq, S., Jaffery, S. H. I., & Khushnood, S. (2014). Effect of current in the EDM machining of aluminum 6061 T6 and its effect on the surface morphology. *Arabian Journal for Science and Engineering*, *39*(5), 4187–4199. https://doi.org/10.1007/s13369-014-1020-z.

Bobbili, R., Madhu, V., & Gogia, A. K. (2015). Modelling and analysis of material removal rate and surface roughness in wire-cut EDM of armour materials. *Engineering Science and Technology, An International Journal*, *18*(4), 664–668. https://doi.org/10.1016/j.jestch.2015.03.014.

Borzíkova, J. (2009). *Analysis of heat-affected zone depth of sample surface at electrical discharge machining with brass wire electrode. Vol. 51* (pp. 633–640). Strojarstvo.

Casati, R., & Vedani, M. (2014). Metal matrix composites reinforced by nano-particles—A review. *Metals*, *4*(1), 65–83. https://doi.org/10.3390/met4010065.

Chaitanya Reddy, M., & Venkata Rao, K. (2020). An overview of major research areas in wire cut EDM on different materials. *INCAS Bulletin*, *12*(4), 33–48. https://doi.org/10.13111/2066-8201.2020.12.4.4.

Choudhary, R., Kumar, H., & Garg, R. K. (2010). Analysis and evaluation of heat affected zones in electric discharge machining of EN-31 die steel. *Indian Journal of Engineering and Materials Science*, *17*(2), 91–98. http://nopr.niscair.res.in/bitstream/123456789/8616/1/IJEMS%2017%282%29%2091-98.pdf.

Cusanelli, G., Hessler-Wyser, A., Bobard, F., Demellayer, R., Perez, R., & Flükiger, R. (2004). Microstructure at submicron scale of the white layer produced by EDM technique. *Journal of Materials Processing Technology, 149*(1–3), 289–295. https://doi.org/10.1016/j.jmatprotec.2003.11.047.

Devarasiddappa, D., & Chandrasekaran, M. (2020). Experimental investigation and optimization of sustainable performance measures during wire-cut EDM of Ti-6Al-4V alloy employing preference-based TLBO algorithm. *Materials and Manufacturing Processes, 35*(11), 1204–1213. https://doi.org/10.1080/10426914.2020.1762211.

Dey, A., Debnath, S., & Pandey, K. M. (2017). Optimization of electrical discharge machining process parameters for Al6061/cenosphere composite using grey-based hybrid approach. *Transactions of Nonferrous Metals Society of China, 27*(5), 998–1010. https://doi.org/10.1016/s1003-6326(17)60117-1.

Garg, R. K., Singh, K. K., Sachdeva, A., Sharma, V. S., Ojha, K., & Singh, S. (2010). Review of research work in sinking EDM and WEDM on metal matrix composite materials. *The International Journal of Advanced Manufacturing Technology, 50*(5–8), 611–624. https://doi.org/10.1007/s00170-010-2534-5.

Hasçalık, A., & Çaydaş, U. (2007). A comparative study of surface integrity of Ti–6Al–4V alloy machined by EDM and AECG. *Journal of Materials Processing Technology, 190*(1–3), 173–180. https://doi.org/10.1016/j.jmatprotec.2007.02.048.

Ho, K. H., & Newman, S. T. (2003). State of the art electrical discharge machining (EDM). *International Journal of Machine Tools and Manufacture, 43*(13), 1287–1300. https://doi.org/10.1016/S0890-6955(03)00162-7.

Kalaiselvan, K., Murugan, N., & Parameswaran, S. (2011). Production and characterization of AA6061–B4C stir cast composite. *Materials & Design, 32*(7), 4004–4009. https://doi.org/10.1016/j.matdes.2011.03.018.

Khan, N. Z., Khan, Z. A., Siddiquee, A. N., & Chanda, A. K. (2014). Investigations on the effect of wire EDM process parameters on surface integrity of HSLA: A multiperformance characteristics optimization. *Production & Manufacturing Research, 2*(1), 501–518. https://doi.org/10.1080/21693277.2014.931261.

Kumar, S., Singh, R., Singh, T. P., & Sethi, B. L. (2009). Surface modification by electrical discharge machining: A review. *Journal of Materials Processing Technology, 209*(8), 3675–3687. https://doi.org/10.1016/j.jmatprotec.2008.09.032.

Kumar, V. M., & Venkatesh, C. V. (2019). A comprehensive review on material selection, processing, characterization and applications of aluminium metal matrix composites. *Materials Research Express, 6*(7). https://doi.org/10.1088/2053-1591/ab0ee3, 072001.

Lau, W. S., & Lee, W. B. (1991). A comparison between edm wire-cut and laser cutting of carbon fibre composite materials. *Materials and Manufacturing Processes, 6*(2), 331–342. https://doi.org/10.1080/10426919108934760.

Lee, L. C., Lim, L. C., Narayanan, V., & Venkatesh, V. C. (1988). Quantification of surface damage of tool steels after EDM. *International Journal of Machine Tools and Manufacture, 28*(4), 359–372. https://doi.org/10.1016/0890-6955(88)90050-8.

Lin, Y.-C., Chen, Y.-F., Lin, C.-T., & Tzeng, H.-J. (2008). Electrical discharge machining (EDM) characteristics associated with electrical discharge energy on machining of cemented tungsten carbide. *Materials and Manufacturing Processes, 23*(4), 391–399. https://doi.org/10.1080/10426910801938577.

Lin, H., Hwang, J.-R., & Fung, C.-P. (2016). Optimization of vacuum brazing process parameters in AA6061 using Taguchi method. *Journal of Advanced Mechanical Design, Systems, and Manufacturing.* https://doi.org/10.1299/jamdsm.2016jamdsm0031, JAMDSM0031.

Liu, J. F., & Guo, Y. B. (2016). Thermal modeling of EDM with progression of massive random electrical discharges. *Procedia Manufacturing, 5*, 495–507. https://doi.org/10.1016/j.promfg.2016.08.041.

Liu, S., Huang, Y., & Li, Y. (2011). A plate capacitor model of the EDM process based on the field emission theory. *International Journal of Machine Tools and Manufacture, 51*(7–8), 653–659. https://doi.org/10.1016/j.ijmachtools.2011.04.002.

Markopoulos, A. P., Papazoglou, E.-L., & Karmiris-Obratański, P. (2020). Experimental study on the influence of machining conditions on the quality of electrical discharge machined surfaces of aluminum alloy Al5052. *Machines, 8*(1), 12. https://doi.org/10.3390/machines8010012.

Ming, W., Zhang, G., Li, H., Guo, J., Zhang, Z., Huang, Y., et al. (2014). A hybrid process model for EDM based on finite-element method and Gaussian process regression. *The International Journal of Advanced Manufacturing Technology, 74*(9–12), 1197–1211. https://doi.org/10.1007/s00170-014-5989-y.

Mohapatra, K. D., Sahoo, S. K., & Bhaumik, M. (2016). Thermal modeling and structural analysis in wire EDM process for a 3D model. *Applied Mechanics and Materials, 279–289.* https://doi.org/10.4028/www.scientific.net/AMM.852.279.

Mohd Abbas, N., Solomon, D. G., & Fuad Bahari, M. (2007). A review on current research trends in electrical discharge machining (EDM). *International Journal of Machine Tools and Manufacture, 47*(7–8), 1214–1228. https://doi.org/10.1016/j.ijmachtools.2006.08.026.

Mouralova, K., Zahradnicek, R., Benes, L., Prokes, T., Hrdy, R., & Fries, J. (2020). Study of micro structural material changes after WEDM based on TEM lamella analysis. *Metals, 10* (7), 949. https://doi.org/10.3390/met10070949.

Mouralova, K., Zahradnicek, R., & Hrdy, R. (2016). Occurrence of globule of debris on surfaces machined by WEDM. *MM Science Journal, 2016*(06), 1630–1633. https://doi.org/10.17973/MMSJ.2016_12_2016200.

Orhadahwe, T. A., Ajide, O. O., Adeleke, A. A., & Ikubanni, P. P. (2020). A review on primary synthesis and secondary treatment of aluminium matrix composites. *Arab Journal of Basic and Applied Sciences, 27*(1), 389–405. https://doi.org/10.1080/25765299.2020.1830529.

Pandey, P. C., & Jilani, S. T. (1986). Plasma channel growth and the resolidified layer in edm. *Precision Engineering, 8*(2), 104–110. https://doi.org/10.1016/0141-6359(86)90093-0.

Patel, J. D., & Maniya, K. D. (2018). A review on: Wire cut electrical discharge machining process for metal matrix composite. *Procedia Manufacturing, 20,* 253–258. https://doi.org/10.1016/j.promfg.2018.02.037.

Patil, C. S., Ansari, M. I., Selvan, R., & Thakur, D. G. (2021). Influence of micro B 4 C ceramic particles addition on mechanical and wear behavior of aerospace grade Al-Li alloy composites. *Sadhana, 46,* 1–9.

Poovazhagan, L., Rajkumar, K., Saravanamuthukumar, P., Javed Syed Ibrahim, S., & Santhosh, S. (2015). Effect of magnesium addition on processing the Al-0.8 mg-0.7 Si/SiC$_p$ metal matrix composites. *Applied Mechanics and Materials, 787,* 553–557. https://doi.org/10.4028/www.scientific.net/amm.787.553.

Pramanik, A., Basak, A. K., Littlefair, G., Debnath, S., Prakash, C., Singh, M. A., et al. (2020). Methods and variables in electrical discharge machining of titanium alloy—A review. *Heliyon, 6*(12). https://doi.org/10.1016/j.heliyon.2020.e05554, e05554.

Pravin, R., & Kavin Raj, S. (2019). Stir casting & processing of aluminum matrix composites. In *Vol. 2142. AIP Conference Proceedings* American Institute of Physics Inc. https://doi.org/10.1063/1.5122395.

Ramanathan, A., Krishnan, P. K., & Muraliraja, R. (2019). A review on the production of metal matrix composites through stir casting—Furnace design, properties, challenges, and research opportunities. *Journal of Manufacturing Processes, 42,* 213–245. https://doi.org/10.1016/j.jmapro.2019.04.017.

Ramnath, B. V., Elanchezhian, C., Annamalai, R. M., Aravind, S., Atreya, T. S. A., Vignesh, V., et al. (2014). Aluminium metal matrix composites—A review. *Reviews on Advanced Materials Science, 38*(1), 55–60. http://www.ipme.ru/e-journals/RAMS/no_13814/06_13814_ramnath.pdf.

Rohatgi, P. (1991). Cast aluminum-matrix composites for automotive applications. *JOM, 43* (4), 10–15. https://doi.org/10.1007/BF03220538.

Sharma, R., Jhap, S., Kakkar, K., Kamboj, A., & Sharma, P. (2017). A review of the aluminium metal matrix composite and its properties. *International Research Journal of Engineering and Technology*, *4*, 832–842.

Sidhu, S. S., Batish, A., & Kumar, S. (2013). Fabrication and electrical discharge machining of metal–matrix composites: A review. *Journal of Reinforced Plastics and Composites*, *32*(17), 1310–1320. https://doi.org/10.1177/0731684413489366.

Singh, B., & Misra, J. P. (2016). A critical review of wire electric discharge machining. *DAAAM International Scientific Book 2016*. Vienna: DAAAM International.

Singh, M. A., Das, K., & Sarma, D. K. (2018). Thermal simulation of machining of alumina with wire electrical discharge machining process using assisting electrode. *Journal of Mechanical Science and Technology*, *32*(1), 333–343. https://doi.org/10.1007/s12206-017-1233-7.

Straka, Ľ., Čorný, I., & Piteľ, J. (2016). Properties evaluation of thin microhardened surface layer of tool steel after wire EDM. *Metals*, *6*(5), 95. https://doi.org/10.3390/met6050095.

Suthar, J., & Patel, K. M. (2018). Processing issues, machining, and applications of aluminum metal matrix composites. *Materials and Manufacturing Processes*, *33*(5), 499–527. https://doi.org/10.1080/10426914.2017.1401713.

Thao, O., & Joshi, S. S. (2008). Analysis of heat affected zone in the micro-electric discharge machining. *International Journal of Manufacturing Technology and Management*, *13*(2–4), 201. https://doi.org/10.1504/IJMTM.2008.016771.

Uthayakumar, M., Babu, K. V., Kumaran, S. T., Kumar, S. S., Jappes, J. T. W., & Rajan, T. P. D. (2019). Study on the machining of Al–SiC functionally graded metal matrix composite using die-sinking EDM. *Particulate Science and Technology*, *37*(1), 103–109. https://doi.org/10.1080/02726351.2017.1346020.

Zhang, C. (2014). Effect of wire electrical discharge machining (WEDM) parameters on surface integrity of nanocomposite ceramics. *Ceramics International*, *40*(7), 9657–9662. https://doi.org/10.1016/j.ceramint.2014.02.046.

Zhang, Y., Liu, Y., Ji, R., & Cai, B. (2011). Study of the recast layer of a surface machined by sinking electrical discharge machining using water-in-oil emulsion as dielectric. *Applied Surface Science*, *257*(14), 5989–5997. https://doi.org/10.1016/j.apsusc.2011.01.083.

CHAPTER EIGHT

Scope of industry 4.0 components in manufacturing SMEs

Arora Monika[a], Buttan Apoorva[b], Kumar Anuj[a], Pujari Purvi[c], and Sabharwal Jyotsana[a]
[a]Apeejay School of Management, Delhi, India
[b]Amity School of Communication, Amity University, Noida, India
[c]Bharati Vidyapeeth's Institute of Management Studies and Research, Mumbai, India

8.1 Introduction

Computational intelligence (CI) is a discipline of technology that incorporates studying the structure of intelligent agents. These agents are triggers within the environment or surroundings like human beings, organizations, bugs, cats, airplanes, and even society. The intelligent agent acts smartly within certain circumstances to achieve the desired goals. It works according to the changing environments and exhibits flexibility while learning from experiences and making appropriate decisions given perpetual roadblocks and finite computation. CI facilitates the understanding of characteristics that make intelligent behavior possible in both natural and artificial systems. Computational intelligence (CI) (Bezdek, 1992) is increasingly being adopted by firms all over the globe to enhance their productivity and market presence and understanding. Computational intelligence is a method through which a system is created, which has a capacity to learn and/or deal in dynamic situations, so that it has the capability to possess one or more attributes of reason, for example, association, generalization, discovery, and abstraction (Eberhart, 2007).

8.2 What is AI?

Artificial intelligence is an extension to human intelligence simulated by computer systems. AI emulates the ability of human brain to study, analyze, learn, and make decisions for complex problems.

For better customer relations, AI can decipher the conversations. It has the ability to analyze interactions on social media and blogs that help in understanding influential markets. SMEs look for little investment in technology but higher value for money. Software as a service (SaaS) provides customized services to SMEs to set up software infrastructure like AI-powered social media strategy.

Components of Artificial Intelligence (Aggarwal, Yadav, Sharma, Uniyal, & Sharma, 2013)

1. The user interface: This lies between the user and the expert system while carrying out a problem-solving process. It aids in confirming the directions to lead the user entry and later interprets the answers which are formed by the system.
2. The information base: It stores major facts about the problem in the background which get built in the system.
3. The shell or interface engine: The interface engine is the program that identifies a suitable information in the information base and receives fresh information streamlining processing and analytical strategies.

8.3 Difference between CI and AI

Computational intelligence is often confused or interchangeably used with artificial intelligence; however, they both are distinct yet similar. As the disciple of AI concentrates on studying the intelligent behavior of the machines and not human beings' natural intelligence, CI, on the other hand, is the study incorporating flexibility of the systems to develop intelligent systems within changing environments. The aim of AI is to develop intelligent systems which showcase intelligent behavior and learn like humans, while the goal of CI is to gauge the computational paradigms existing to enable intelligent behavior in both natural and artificial systems within complex and changing surroundings.

Some of the applications of AI used commonly are speech recognition, handwriting recognition, character recognition, machine vision, big data solutions, and many more. Some applications of CI much used are smart household appliances, banking optimization, industrial applications, medical diagnosis, and so on (Khillar, 2020).

8.4 Principles of CI

Fuzzy logic exhibits the qualitative and inconsistent approach in human reasoning, which assists the robust systems. The fuzzy logic algorithm is smart and considers several possibilities of a problem and not only True or False. The concept of fuzzy logic was proposed by Lotfi Zadeh in

1965. The computer's ability to work like humans and assume possible values between True and False. The algorithm can be used for vague concepts where the truth degree ranges between 0.0 and 1.0. The high acceptance of fuzzy logic systems in many industries because it enables reprogramming in the absence of feedback sensors.

Fuzzy logic is basically the first-order logic, which works on binary where the truth of a statement is represented between the values of 0 and 1 where False (0) and True (1). Fuzzy logic is now incorporated in several fields like medicine, transportation systems, defense, and manufacturing.

8.5 Neural network

It mimics the way neural networks in the human brain work (Hansen & Bøgh, 2021). Just like in biological neural network where the neurons receive multiple input signals and identify each signal based on the importance to be assigned to each response, the artificial neural network also works in the same fashion. They get trained over time through multiple rounds of trial and error just like brain. Neural networks are generally aligned in layers, which are made of numerous interconnected nodes (Aggarwal et al., 2013).

8.6 Deep learning

It is a subset of machine learning, which falls under artificial intelligence. Deep learning uses neural networks. The neural network gets trained from the enormous data available. The availability of immense data sets and the need to analyze this data is becoming more pertinent creating new algorithms based on deep learning models.

Reinforcement learning is the training of machine learning models to make a sequence of decisions. The agents are taught to be goal-directed in uncertain and complex circumstances. During such learning method, AI deploys trial-and-error problems to come up with the most appropriate solutions. To enable the machine to act in accordance with the programmer, artificial intelligence would get appreciated or penalized for its action. It's a game-like structure where AI is rewarded.

8.7 Probabilistic method for uncertain reasoning

Many-a-times the agents have to deal with uncertain or incomplete datasets. The researchers have devised Bayesian networks which are brought to use while handling innumerable problems at once. The probabilistic algorithms are used for filtering, prediction, smoothing, and searching explanations for several domains of data, which help the flexible perception systems to analyze processes that occur concurrently. A mathematical tool has been developed with precision to understand the decision making of agents using multiple tools to aid better productivity and avoid errors.

8.8 Small and medium enterprises (SMEs)

There have been numerous attempts at defining SMEs. All over the globe, different economies have defined SMEs on different parameters ranging from number of employees to turnover to investment. The World Bank defines SMEs on the basis of strength of employees, i.e., microscale; less than 50 employee, small scale; 50 employees, medium scale; 50–200 employees. The Government of India classified Micro, Small, Medium Manufacturing Enterprises and Enterprises rendering Services by Investment in Plant and Machinery or Equipment. For micro, it is not more than Rs. 1 crore and annual turnover, not more than Rs. 5 crores; for small enterprises, it is not more than Rs. 10 crore and annual turnover, not more than Rs. 50 crores; and for medium, it is categorized as not more than Rs. 50 crores and annual turnover, not more than Rs. 250 crores, applicable from July 2020 (Source: https://msme.gov.in/know-about-msme).

Whichever the definition is in use, the role of SMEs in growth of any economy cannot be denied. The converging point in all the literature is that they play a crucial role in the development of any country. All the developing economies need these enterprises to ensure a balanced regional development (Sharma, Panthey, Kumar, & Kour, 2014). Recently, SMEs have been under various pressures and challenges due to the pandemic COVID-19. SMEs have been facing multiple issues on both demand and supply sides of the business, post-COVID-19. On one side, they are facing difficulty in continuing with production due to labor and raw material shortage. On the other side, the consumer demand and consumer reach has been curtailed due to the pandemic (Kumar & Ayedee, 2021).

8.9 Industry 4.0

Industry 4.0 is a broad domain including production systems, customer relationship, and data management, etc. (Segovia, Mendoza, Mendoza, & González, 2015). Industry 4.0 is considered as a comprehensive name for a cluster of inter-related technologies, which have had a significant impact on enhanced pace of digitization of the firms through innovations and growth (Burritt & Christ, 2016). These cluster of technologies have been majorly identified as artificial intelligence, machine learning, Internet of things (IoT), Big Data analytics, and augmented reality (AR)—virtual reality (VR). The other terms that are used for the same are smart manufacturing, cloud manufacturing, intelligent manufacturing, fourth industrial revolution, etc. (Culot, Nassimbeni, Orzes, & Sartor, 2020).

8.10 SMEs and industry 4.0

The significance of small and medium enterprises in the balanced and sustainable development of an emerging economy like India is paramount. Indian SMEs have long faced the challenge of being technologically disadvantaged compared to SMEs of other countries as well as MNCs. Researchers all over the globe have been reiterating the need of better technology adoption by SMEs. Pre-COVID-19 and during the COVID-19, it has been evident that the firms that ensure extensive adoption of technologies in daily operations are much more successful. There are different components of technology that an organization can adopt. Those components include e-commerce, social media, ICT tools, mobile commerce platforms, artificial intelligence, cloud computing, block chain, robotics, etc. It is not a choice for SMEs to adopt technology, especially after COVID-19; it is mandatory for them to go for adoption. Most of the researchers conclude that the use of technologies will become litmus test of survival of SMEs. Artificial Intelligence is one of the most important technology in domain of Industry 4.0. Adoption of AI has become imperative for SMEs to enhance their reach to customers and improve efficiency (Kumar & Ayedee, 2021).

8.11 Artificial intelligence

There have been numerous studies reflecting the benefits accruing to SMEs on account of usage of AI in social media–assisted marketing (Basri, 2020a, 2020b). AI has been found to be a wonderful tool in enhancing productive efficiency, competitive advantage, and customer engagement and market penetration for SMEs (Table 8.1).

Table 8.1 Review of literature.

S. no	Author	Adoption field by SMEs
1.	Basri (2020a, 2020b)	This paper dwells on adoption of AI by Saudi Arabia SMEs specifically in the field of social media-assisted marketing. The paper concludes that there has been a clear increase in productivity, workforce performance, and customer bases through the usage of AI
2.	Hansen and Bøgh (2021)	The paper emphasizes the need of adoption of AI for the SMEs to stay competitive. It also suggests using IoT for other inputs for, e.g., machine-wise predictive analytics
3.	Akpan, Udoh, and Adebisi (2020)	The authors argue that disruptive computing technologies, data analytics, and the Internet of things (IoT) required to engineer new business models, reduce overheads, enhance competitive advantages, and digitize SMEs' business operations remain untapped in developing economies
4.	Ghobakhloo and Ching (2019)	The authors suggest that technological, organizational, and environmental factors determine SMEs' decision to adopt digital technologies like AI

8.12 Barriers to adoption of computational intelligence in SMEs

SMEs must overcome the technical and attitudinal barriers in adoption of computational intelligence. They have to ensure the technical infrastructure and availability of well-managed data to enable the adoption. SMEs account for approximately 99% of total firms in the nation, creating employment for more than 50% of private sector employees and creating almost more than 50% of the nation's nonfarm GDP (Brief, 2000).

8.13 Computational intelligence and SMEs

The SMEs can benefit from the computational intelligence approach where the SMEs can be using their transactional data either in the form of ERP or in the form of MIS, where the departments are connected with the different functions across the departments. Also, the use of computers helps to improvise the processes and also increase the efficiency and performance.

This also helps the organization to grow and be better. This will be helping to improve the economy to grow and will be better for sure. The collected data will be helping in getting the information in terms of reports etc. to be secure for the future. Also, this information will be helpful and giving input to artificial intelligence, and it will be helpful in creating and be a part of hard computing. Also, data uses fuzzy logic, genetic algorithm, and artificial neural networks. That gives the input to soft computing and further helps in creating computational intelligence. Computational intelligence involves reinforcement learning. Reinforcement learning collective human knowledge which makes the better decisions. Together this will help in achieving machine intelligence. This claims that soft computing is a subset of computational intelligence and computational intelligence is a subset of artificial intelligence.

The SMEs use the computational intelligence along with the probabilistic feedback and take the decision for justification of decision. SMEs with computational intelligence will always be performing the correct decision as it was based on reinforcement learning as shown in Fig. 8.1, where SMEs respond back and will decide on the computational intelligence and artificial intelligence will decide for using soft and hard computing. The different techniques are used to understand the data used in different processes. The techniques used have sense of data used in the processes and works for the customer satisfaction. The response for the users is generated using

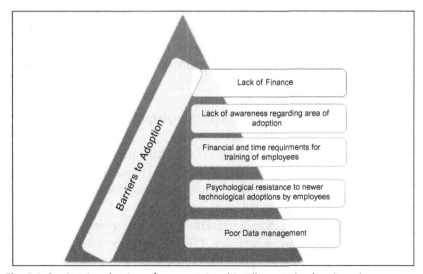

Fig. 8.1 Barriers in adoption of computational intelligence (authors' own).

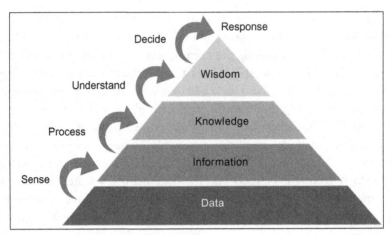

Fig. 8.2 Intelligence layers (authors' own).

the technological innovation. It helps the user to work efficiency and effectively (Fig. 8.2).

As the SMEs are growing at a rapid pace, it becomes important to address the need for and importance of culminating the different layers to form a structure to serve the customers and their needs. Data layer differs from the information and intelligence layers at large.

Data layer—The data or the technology and artificial intelligence associated with SMEs runs the entire sector. This has resulted in the betterment of the sector by providing cost reductions that will help to match the SMEs with bigger players operating in the market.

Artificial intelligence and the data gathered from it have resulted in
1. Proper establishment of CRM
2. Sales management
3. Knowing and analyzing the needs and preferences of the customers
4. Safeguarding against cybercrimes, etc.

Thus, the coming up of AI has enabled the SMEs to use data-driven tools that help in knowing the analytics of customer preferences. Also, AI helps in tapping and gathering of the new domains which help the small manufacturing units to have an edge ahead of its competitors.

Information layer—It defines the skills, capabilities, traits, and characteristics that the human resources possess. It is important to understand that for SMEs to grow up to their full potential, they still need to abide by the information and work of the human assets. Though technology helps in creating a smooth way for the firms to carry out their processes, it's the human resources that monitor its way.

Human Resource Management becomes an integral part for the SMSEs because it helps in

a. Employee engagement

b. Maintaining quality of work

c. Easing out the delivery mechanism

d. Integrating artificial intelligence with humans

e. Enhancing productivity, etc.

Intelligence layer—It involves the use of "intelligence," that is, it includes the role of both the human resources and the artificial intelligence. It involves some major components to focus upon-.

CRM—customer relationship management

The integration of artificial intelligence with the human resources has resulted in a tremendous amount of growth for the organizations.

Artificial intelligence and human knowledge together have created wonders. The coming up of chatbots has resulted in building up better customer relationships.

Business processes

While human intelligence tends to adapt new skills and capabilities, artificial intelligence focuses on the latest technology that may try to mimic human capabilities. Thus, these two when put together try to achieve, develop, and add more to the business processes.

Thus, artificial intelligence is supporting the human assets to escalate their cognitive strengths, skills, and abilities. Their amalgamation has resulted in a thumbs-up for the businesses to proceed further.

8.14 Model formulation (data layer/information layer/intelligence layer)

See Table 8.2.

8.15 Model discussion

Artificial intelligence has grabbed the market, and also the computational intelligence has a way forward to go. Artificial intelligence includes expert's system, agent, natural language processing, robot, machine vision, and speech recognition. The data patterns used in AI will be helpful for achieving the computational intelligence by using probabilistic reasoning

Table 8.2 Layer-wise description of computational and artificial intelligence.

Layer	Description	Computational intelligence	Artificial intelligence
Data layer	Techniques used	Soft computing techniques	Hard computing techniques
Intelligence layer	Applications	Evolution programming, genetic algorithm	Manufacturing, automation, robotics, transport, and healthcare
Data layer	Logic	Follows fuzzy logic	Follow binary logic
Information layer	Models	Nature-inspired models	Based on mathematical models
Data layer	Nature of data	Can work inexact and incomplete data	Not very effective
Intelligence layer	Output/ results	Probabilistic results	Deterministic results

and belief network. The computational intelligence uses fuzzy logic, neural network and genetic algorithm, swarm intelligence, and artificial life.

8.16 Applications of CI in SMEs

The coming up of artificial intelligence has paved the way for the bigger industries as well as the knowledge, skills of the employees associated with it, but when it comes to understanding the same in SMEs, there's a lot to decide.

Thus, there are many benefits associated with the linkage of computational intelligence to SMEs

1. Computational intelligence intensifies marketing—The benefits of computational intelligence have added a lot to the field of marketing. Now, the automation system has not only helped the clients to reach the market more easily, but it has also helped the employees, the human resources of an organization to easily track the customers' spends, and how they react to a particular product. CI has helped to mold its way toward direct marketing that empowers the databases marketing processes toward computational intelligence.

2. Computational intelligence drives in with CRM—If the CRM (customer relationship management) drives in with the computational intelligence, it may drive the integration of customer relationships with CI on another platform. It helps to judge the consumers' sentiments via social

media posts, and helps to record conversions, customers' feedback, and reviews. Also, it helps to find out the results using data analysis.

3. The establishment of CI chatbots—When talking about the relationship of artificial intelligence with human intelligence, the truest examples for the same are the chatbots. These are now used by many organizations and businesses to ensure that their audience and customers are engaged. With the help of these chatbots, the businesses tend to address the customer's needs, problems, and queries 24/7. Thus, the integration of AI and CI tends to create a feeling among the customers that they are important. Thus, it paves the way for more technological advancements in the near future.

4. CI reinforces cybersecurity—Cybersecurity has been one of the biggest threats to businesses operating majorly on social media platforms. Thus, in order to overcome this hindrance, CI has eased the way for businesses in spotting strange behaviors and vulnerabilities. Though small enterprises are more vulnerable to cybersecurity attacks, measures are being built for the same.

5. CI supports competitive intelligence—Computational intelligence helps in determining the competitor's actions well in advance. The tools utilized under CI help in impressive analytical skills. It allows the businesses to dominate the PR activities. Following this, the firms can improve their strategies and set their goals accordingly.

8.17 Conclusion

The study was carried out for SMEs, where the researchers have identified that there can be a possibility of technological adoption. The technological adoption will not only help the SMEs to grow and perform better but also increase their efficiency by the incorporation of computational intelligence. The concept of computation intelligence is emerging these days where it takes cares of artificial intelligence and machine learning. It was based on soft computing where the knowledge of human is also imbibed and has helped in the creation of a probabilistic way to process the activities. Computational intelligence is a subset of artificial intelligence, which works on technological innovation. The machine learning works on the data and its use in the computation of models, which will be helpful for machines to work with more intelligently and accurately.

The implication for SMEs depends on artificial and human intelligence. The coming up of computational intelligence, and its amalgamation with artificial and human intelligence, has resulted in a variety of new prospects for the development of SMEs.

It has resulted in the following:

1. It enhances the efficiency and effectiveness of the working of SMEs
CI has substantially reduced red tapism, has helped in skill management and job evaluation, and has helped in a digital transformation, thus increasing the efficiency and effectiveness of the working of the SMEs.

2. It has reduced the cost of experimentation and innovation
Experimenting has become much easier with the introduction of artificial intelligence. It has helped to pave the way by coming up with new ideas and using them to check for the deviations. Thus, it helps SMEs to try out new things that work in its favor.

3. Data is the key and human factor is critical
To understand and work with the technology, it becomes essential to value the human factors associated with it. After all, humans have created the technology.

Thus, for SMEs, both the data system and the human factor play an important role.

Implication for economy as the use of AI and CI has been affected the growth of SMEs. Many renowned researchers have predicted that artificial intelligence will transform the economy majorly by 2030. The major reason behind this is that CI helps in productivity enhancement, helps in employment, and also helps in contributing to the GDP. Also, CI helps in taking up risky tasks that, when worked in favor, can contribute toward the development of the economy. The CI has the ability to create new technologies and products; thus when these products are presented in the market, it can result in the economy to boost.

The future aspects of computational intelligence in SMEs are vast. According to the Mckinsey Report, it states that by 2030, the CI will capture around 16% increases in the economy. Thus, when talking about its development in SMEs, it can be incurred that

1. It will enhance the cyber security system—It is estimated that CI has the capability of enhancing cyber security, i.e., it will add more protection to the data available on the World Wide Web. This will help the small businesses to grow more safely and securely.

2. CI will help to manage data analytics—The technology associated with CI helps to gain insights; the coming up of latest platforms such as Google Analytics, automation platforms, CRMs, etc., will help the SMEs to not only manage its data effectively but to use the data as per their need and convenience.

3. It will help to amplify the current capital capacity—When talking about the CI's ability to amplify the capital, it simply means that it has the power to bring in the change. This means that it can very accurately determine the estimation of the cost required to form capital. Not only that, CI brings in new methods to estimate the requirements of the how and when the funds are needed.

4. CI will help in the examination of problems more effectively. The increasing pace at which the technology is growing estimate that CI has the power to open up new dimensions.

It is estimated that in the upcoming years, CI will be more advanced and will have a much better capability to estimate, record, and predict things more accurately.

This will help the SMEs to ease their procedures and will help them to achieve more.

References

Aggarwal, P., Yadav, P., Sharma, N., Uniyal, R., & Sharma, S. (2013). Research paper on artificial intelligence. *Case Studies Journal*, 7–14.

Akpan, I. J., Udoh, E. A. P., & Adebisi, B. (2020). Small business awareness and adoption of state-of-the-art technologies in emerging and developing markets, and lessons from the COVID-19 pandemic. *Journal of Small Business & Entrepreneurship*, 1–18.

Basri, W. (2020a). Examining the impact of artificial intelligence (AI)-assisted social media. *International Journal of Computational Intelligence Systems*, 13(1), 142–152.

Basri, W. (2020b). Examining the impact of artificial intelligence (Ai)-assisted social media marketing on the performance of small and medium enterprises: Toward effective business management in the Saudi Arabian context. *International Journal of Computational Intelligence Systems*, 13(1), 142–152. https://doi.org/10.2991/ijcis.d.200127.002.

Bezdek, J. (1992). On the relationship between neural networks, pattern recognition and intelligence. *International Journal of Approximate Reasoning*, 6(2), 85–107.

Brief, O. P. (2000). *Small and medium-sized enterprises: Local strength, global reach, s.l.* Public Affairs Division, OECD.

Burritt, R., & Christ, K. (2016). Industry 4.0 and environmental accounting: a new revolution? *Asian Journal of Sustainability and Social Responsibility*, 1(1), 23–38.

Culot, G., Nassimbeni, G., Orzes, G., & Sartor, M. (2020). Behind the definition of Industry 4.0: Analysis and open questions. *International Journal of Production Economics*, 226, 107617.

Eberhart, R. C. (2007). *Computational intelligence: Concept to implementations*. Morgan.

Ghobakhloo, M., & Ching, N. T. (2019). Adoption of digital technologies of smart manufacturing in SMEs. *Journal of Industrial Information Integration*, 16.

Hansen, E. B., & Bøgh, S. (2021). Artificial intelligence and internet of things in small and medium-sized enterprises. *Journal of Manufacturing Systems*, 58, 362–372.

Khillar, S. (2020). *Difference between AI and CI*. Retrieved from Difference Between Similar Terms and Objects http://www.differencebetween.net/technology/difference-between-ai-and-ci/.

Kumar, A., & Ayedee, N. (2021). Technology adoption: A solution for SMEs to overcome problems during COVID-19. *Academy of Marketing Studies Journal*, 25(1).

Segovia, D., Mendoza, M., Mendoza, E., & González, E. (2015). Augmented reality as a tool for production and quality monitoring. *Procedia Computer Science, 75,* 291–300. https://doi.org/10.1016/j.procs.2015.12.250.

Sharma, M., Panthey, R., Kumar, R., & Kour, G. (2014). Role of SMES in India economy and TQM. *International Journal of Business Management, 1,* 119–128.

CHAPTER NINE

Process parameter optimization in manufacturing of root canal device using gorilla troops optimization algorithm

Swanand Pachpore[a,b], Pradeep Jadhav[b], and Ratnakar Ghorpade[a]
[a]School of Mechanical Engineering, Dr. Vishwanath Karad MIT World Peace University, Pune, India
[b]Bharati Vidyapeeth (Deemed to be University) College of Engineering, Mechanical Engineering Department, Pune, India

9.1 Introduction

Endodontology is a field of study that deals with the knowledge of what is within the tooth and that deals with thorough chemo-mechanical cleaning of the root canal system. A root canal is a dental operation that involves the removal of the pulp (root nerves and blood cells) from a tooth, cleaning and sculpting the canal cavity, and then introducing fillers to prevent germs from re-entering the root canal's apical end. Cleaning and shaping the canal cavity is referred to as "biomedical preparation," and 3D filling and sealing the canal to establish an undamaged seal is referred to as obturation. In other words, obturation is the process of filling the root canal cavity with heated and softened gutta-percha to establish a fluid-tight seal between the canal wall and the filled gutta-percha. Basically, gutta-percha is a natural isoprene polymer extracted from the resin and sap of the trees of the Palaquium family. In a recent study, the researcher has proposed that during filling, softened gutta-percha is pushed into the canal, creating a high frictional force, which is nothing but resistance offered by the canal wall, which is directly proportional to the smoothness of the canal's curvature after biomechanical preparation (Pachpore, Jadhav, & Ghorpade, 2021). The importance of geometrical parameters is governed by researchers stating that geometrical parameters play the most important role in structuring the canal and establishing the logical cavity volume (Lokhande & Balaguru, 2020).During the process of obturation, softened gutta-percha is subject to condensation done by various methods, which result in the generation of forces classified as lateral forces, which tend to push filling materials against canal walls and into lateral canals, and vertical pressures, which tend to push filling materials in an apical direction, increasing the danger of extrusion. In his recent study, one of the researchers clearly mentioned that the success index of treatment largely depends on the interaction between process parameters, geometry parameters, energy parameters, and material parameters. Assuming material parameters are constant and others are variable, researchers stated that one of the geometry parameters, i.e., curvature coefficient (C), plays an important role in estimating interaction between process parameter, i.e., condensation force, and geometry parameters, i.e., curvature coefficient, and it was found that efficacy of the treatment will be higher for the canal having less curvature than for the canal having higher curvature (Pachpore et al., 2021). The presented work deals with the estimation of sliding friction using the theoretical model and optimizing it to understand the nature of the same along the canal length with respect to vertical condensation. Basically, condensation force optimization is dependent on a precise balance between the taper and diameter of the canal, master cone, and condensation plugger, which needs to evaluate.

This paper provides the application of the Gorilla Troops optimization algorithm (Abdollahzadeh, Soleimanian Gharehchopogh, & Mirjalili, 2021) in real-life problems. Nowadays bio-inspired, nature-inspired algorithms and socio-inspired algorithms play a very vital role in solving real-life problems. Many such bio-inspired, nature-inspired algorithms and socio-inspired algorithms (Patel, Kakandikar, & Kulkarni, 2020), such as genetic algorithm (Bhoskar et al., 2015), particle swarm optimization (Kulkarni et al., 2015), cuckoo search optimization (Joshi, Kulkarni, Kakandikar, & Nandedkar, 2017), bat algorithm (Burande, Kulkarni, Jawade, & Kakandikar, 2021), grasshopper optimization algorithm (Neve, Kakandikar, Kulkarni, & Nandedkar, 2020), firefly algorithm (Kakandikar, Kulkarni, Patekar, & Bhoskar, 2020), flower pollination algorithm (Yang, 2014), cohort intelligence algorithm (Kulkarni, Kulkarni, Kulkarni, & Kakandikar, 2017), etc., have always proven their application in real-life mechanical problems, and results are always outperforming within. With the same reference for when applied for biomedical engineering, typically to understand sliding friction in obturation, Gorilla optimization algorithm is used and evaluated in this research.

9.2 Mathematical model

The resistance to motion of one item moving relative to another is defined as friction. It is not a basic force in the same way that gravity or electromagnetism is. Scientists believe it is caused by the electromagnetic attraction of charged particles on two surfaces that are in contact. In root canal treatment, the softened gutta-percha on the application of condensation technique is pushed inside the canal causing sliding friction. This frictional component will act/evolve along the surface of canals as shown in Fig. 9.1 (Pachpore et al., 2021).

In a recent study, the researcher stated based on geometrical parameters and the frictional coefficient of the biomaterial, the directions and amplitude of frictional force are measured. Moreover, categorizing the canal eases

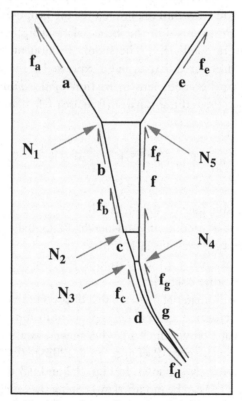

Fig. 9.1 Friction model for biomechanically prepared canal.

the determination of the smoothness of frictional force. The researcher states that typical canal connections are found to be linear in coronal to middle and nonlinear in middle to apical sections, and the linear curvature connectivity results in a low-stress field and no microfracture wherein nonlinear curvature connectivity experiences large stress fields and microfractures with the increase in condensation force. As a result, the outcome of obturation is primarily dependent on biomedical preparation, which is considered to be based on the degree of smoothness of canal walls. Thus, the estimation of sliding friction was very much related to geometric parameters, viz., Taper, relative angle for orientation of cross section between two planes, pitch and polar symmetry constant of the endodontic file stated by researchers in their recent studies (Lokhande & Balaguru, 2020). As stated when softened gutta-percha was forced inside the cavity, condensation force is nothing but the sum of frictional force, hydrodynamic effect, and weight of compacting mass; components of static force and dynamism are involved within the process. In the proposed research work, an attempt is made to establish the exact relationship between frictional force and curvature coefficient. The sliding friction model proposed by the researcher clearly stated that it can be expressed as the sum of the compressive load (F_P), the coefficient of friction of the biomaterial (μ), the working pressure (P), and the canal surface area (A_C), and is expressed as in Eq. (9.1):

$$F_R = F_P + (\mu * P * A_C) * (\cos \theta + \sin \theta) \qquad (9.1)$$

wherein,

F_P = Compressive load (2.495–4.124N)

μ = Coefficient of friction of gutta-percha (0.45–0.49)

P = Condensation pressure in N/mm^2

A_C = Surface of canal = $\pi * d * l$ in mm^2

θ = Canal curvature coefficient

When viscoelastic biomaterial (material that displays both elastic and viscous behaviors when deformed under time-dependent loading circumstances) is forced into a canal cavity, which may be viewed as a microchannel, the hydrodynamic effect occurs (Pachpore et al., 2021). If the canal sections are analyzed individually as shown in Figs. 9.2 and 9.3 to establish sliding friction, then Eq. (9.1) can be modified by keeping the canal length the same as 5 mm in each section.

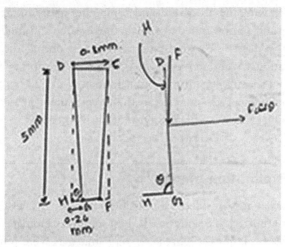

Fig. 9.2 Sliding friction in middle ⅓rd of the canal.

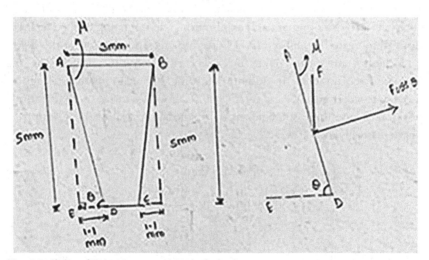

Fig. 9.3 Sliding friction in coronal ⅓rd of canal.

9.3 Gorilla troops optimization for sliding frictional force

Gorilla Troops optimization (GTO) is based on the behavior of gorilla groups that mimic five strategies. These strategies consist of traveling to an undiscovered area, moving to other gorillas, moving in the indicated direction, following the silverback, and competing with adult females. They are simulated and demonstrated to explain the research and application of the optimization process. Three methods were used during the exploration phase: migration to undiscovered areas, migration to other gorillas, and migration to a specific location (Abdollahzadeh et al., 2021; Ginidi et al., 2021). Two strategies are pursued in the development phase: keeping an eye on the silver medal and fighting for adult women.

9.4 Exploration phase

All gorillas are considered as possible solutions in GTO, and at each stage of the optimization operation, the best solution candidate is considered to be the silverback gorilla. Three different strategies were pursued during the exploration phase, namely, migrating to unknown locations to increase the scope of GTO research, migrating to other gorillas to improve the balance between development and exploration, and migrating to specific locations to increase the ability of GTO to find any optimization of the field.

When r and $<$ parameter (p), select the transition strategy to the unrecognized position. In addition, if rand > 0.5, choose the strategy of moving to other gorillas, and if rand < 0.5, choose to move to the identified position. These three strategies in the exploration phase can be expressed mathematically as follows:

$$GXt+1 = \begin{cases} UL - LL \times r_1 + LL, & \text{rand} < p; \\ r_2 - C \times X_r t + L \times H, & \text{rand} \geq 0.5 \\ Xi - L \times L \times Xt - GX_r t + r_3 \times Xt - GX_r t, & \text{rand} < 0.5 \end{cases}$$

$$(9.2)$$

"where $X(t)$ and $GX(t+1)$ represent the current vector of gorilla position and the candidate position vector of the gorilla in the following t iteration, respectively, while rand, r_2, r_2, and r_3 signify the random values in the range from 0 to 1. The parameter (p) demonstrates the probability of choosing the migration strategy to an unidentified position and must be specified in a range of 0–1 before the optimization operation. The parameters Xr and GXr illustrate one member of the gorillas designated from the whole population and one of the vectors of gorilla candidate positions that can be randomly designated, respectively. LL and UL characterize the lower and upper limits of the variables, respectively. The variables C, L, and H can be mathematically represented according to Eqs. (9.4), (9.6), and (9.7), respectively" (Ginidi et al., 2021).

$$C = F \times (1 - It/Maxit), \tag{9.3}$$

$$F = cos(2 \times r_4) + 1, \tag{9.4}$$

$$L = C \times l, \tag{9.5}$$

$$H = Z \times X(t), \tag{9.6}$$

$$Z = [-C, C] \tag{9.7}$$

The symbols It and $MaxIt$ represent the total value of the current iteration and optimization process individually, and the symbols cos and r_4 refer to the random values in the range of 0–1, respectively. In addition, the characters l and Z represent the random values in the range of $[-1.1]$ and $[-C, C]$, respectively. Estimate the cost of all solutions GX at the end of the exploration phase. If the cost is $GX(t) < X(t)$, then the solution $GX(t)$ will replace the solution $X(t)$ and become the best solution (silverback).

9.5 Exploitation phase

"During the development phase of GTO, two strategies were adopted: following the silverback and competing for adult females. By using the value of C in Eq. (9.4) and comparing it with parameter W (which can be set), one of two strategies can be selected, as shown in the next section. The silver gorilla is the leader of a group. It makes decisions and guides the other gorillas toward food sources. This behavior can be expressed mathematically according to Eq. (9.8)" (Ginidi et al., 2021).

$$GX(t + 1) = L \times M \times (X(t) - X_{siverback}) + X(t) \tag{9.8}$$

$X(t)$ signifies the gorilla position vector, while $X_{siverback}$ signifies the silverback gorilla position vector, which offers the best solution.

$$M = \left(\left| (1/N) \sum_{i=1}^{N} GX_i(t) \right|^g \right)^{(1/g)} \tag{9.9}$$

$GXi(t)$ shows the location of each candidate gorilla's vector in iteration t, whereas N signifies the number of gorillas.

$$g = 2^L \tag{9.10}$$

L can be determined by Eq. (9.5).

"The competition of adult females is the second strategy designed for the development stage if $C < W$. When young gorillas mature, they compete fiercely with other males for the choice of adult females. This behavior can be expressed mathematically according to Eq. (9.11)" (Ginidi et al., 2021).

$$GX(i) = X_{siverback} - (X_{siverback} \times Q - X(t) \times Q) \times A, \qquad (9.11)$$

$$Q = 2 \times r_5 - 1 \qquad (9.12)$$

$$A = \beta \times E \qquad (9.13)$$

$$E = \begin{cases} N_1 \text{rand} \geq 0.5 \\ N_2 \text{rand} < 0.5 \end{cases} \qquad (9.14)$$

"Q simulates the impact force, expressed in Eq. (9.12), and the symbol r_5 shows a random value in the range of [0,1]. The coefficient A is a vector representing the degree of violence in the event of a conflict, which can be estimated using Eq. (9.13). In Eq. (9.13), parameter b is the target value before the optimization operation, and E is used to simulate the impact of violence on the decision size.

The cost of all GX solutions is estimated at the end of the run phase. If the cost of $GX(t) < X(t)$, *the* $GX(t)$ solution $GX(t)$ will replace the solution $X(t)$ and become the best solution (silver back). Fig. 9.4 depicts the main stages of GTO development for extracting solar cell model parameters" (Ginidi et al., 2021).

9.6 Results and discussion

The frictional force estimation is carried out in each part of the canal with equations stated above, which clearly indicates that in all the sections, the frictional force is directly proportional to the product of condensation force expressed in terms of curvature coefficient, which is geometry parameter, and coefficient of friction of gutta–percha, which is a material parameter; only magnitude differs. In the coronal section, the relation was found to be $0.2\pi F$ and in the middle section was found to be $0.05\pi F$. And in the apical region, the value was very negligible. Table 9.1 shows the comparison between values of frictional force obtained through analytical method and optimizer.

The table clearly shows that frictional force constitutes 1/3rd of the total condensation force. In order to achieve 100% success of treatment, the other

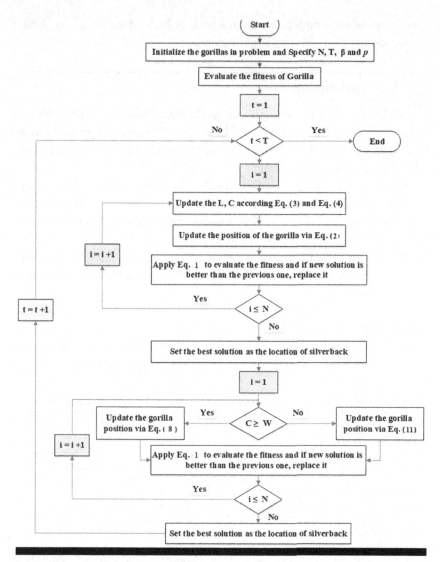

Fig. 9.4 Flowchart of GTO. *From Ginidi, A., Ghoneim, S. M., Elsayed, A., El-Sehiemy, R., Shaheen, A., & El-Fergany, A. (2021). Gorilla troops optimizer for electrically based single and double-diode models of solar photovoltaic systems. Sustainability, 13(16), 9459. https://doi.org/10.3390/su13169459.*

Table 9.1 Result comparison for frictional force.

Sr. no	Region	Analytical method	GTO
1	Coronal 1/3rd	3–8 N	7.7302
2	Middle 1/3rd	2–6.5 N	5.4332
3	Apical 1/3rd	0.6–3 N	3.9351

parameters of the condensation force such as hydrodynamic effect during obturation need to estimate experimentally.

9.7 Conclusion

The comparative study performed clearly indicates that sliding friction plays a very crucial role in enhancing the success index of root canal treatment, i.e., obturation. This leads to points that need to be taken into consideration as mentioned below.

The analytical estimation of frictional force clearly indicates that the frictional force with in the coronal region was found to be 20% of the total condensation force and in the middle region, the value ranges approximately within 5% of the total condensation force.

This clearly indicates that as we move toward the apical region, the magnitude of sliding friction falls, which resembles the assumptions of the traditional treatment.

References

Abdollahzadeh, B., Soleimanian Gharehchopogh, F., & Mirjalili, S. (2021). Artificial gorilla troops optimizer: A new nature-inspired metaheuristic algorithm for global optimization problems. *International Journal of Intelligent Systems*, *36*(10), 5887–5958. https://doi.org/10.1002/int.22535.

Bhoskar, T., Kulkarni, O. K., Kulkarni, N. K., Patekar, S. L., Kakandikar, G. M., & Nandedkar, V. M. (2015). Genetic algorithm and its applications to mechanical engineering: A review. *Materials Today: Proceedings*, *2*(4–5), 2624–2630. https://doi.org/10.1016/j.matpr.2015.07.219.

Burande, C. G., Kulkarni, O. K., Jawade, S., & Kakandikar, G. M. (2021). Process parameters optimization by bat inspired algorithm of CNC turning on EN8 steel for prediction of surface roughness. *Journal of Mechatronics and Artificial Intelligence in Engineering*. https://doi.org/10.21595/jmai.2021.22148.

Ginidi, A., Ghoneim, S. M., Elsayed, A., El-Sehiemy, R., Shaheen, A., & El-Fergany, A. (2021). Gorilla troops optimizer for electrically based single and double-diode models of solar photovoltaic systems. *Sustainability*, *13*(16), 9459. https://doi.org/10.3390/su13169459.

Joshi, A. S., Kulkarni, O., Kakandikar, G. M., & Nandedkar, V. M. (2017). Cuckoo search optimization—A review. *Materials Today: Proceedings*, *4*(8), 7262–7269. https://doi.org/10.1016/j.matpr.2017.07.055.

Kakandikar, G. M., Kulkarni, O., Patekar, S., & Bhoskar, T. (2020). Optimising fracture in automotive tail cap by firefly algorithm. *International Journal of Swarm Intelligence*, *5*(1), 136. https://doi.org/10.1504/ijsi.2020.106396.

Kulkarni, O., Kulkarni, N., Kulkarni, A. J., & Kakandikar, G. (2017). Constrained cohort intelligence using static and dynamic penalty function approach for mechanical components design. *International Journal of Parallel, Emergent and Distributed Systems*, *33*(6), 570–588. https://doi.org/10.1080/17445760.2016.1242728.

Kulkarni, N. K., Patekar, S., Bhoskar, T., Kulkarni, O., Kakandikar, G. M., & Nandedkar, V. M. (2015). Particle swarm optimization applications to mechanical engineering—A review. *Materials Today: Proceedings*, *2*(4–5), 2631–2639. https://doi.org/10.1016/j.matpr.2015.07.223.

Lokhande, P. R., & Balaguru, S. (2020). A mathematical model for root canal preparation using endodontic file. *Journal of Oral Biology and Craniofacial Research*, *10*(1), 396–400. https://doi.org/10.1016/j.jobcr.2019.09.002.

Neve, A. G., Kakandikar, G. M., Kulkarni, O., & Nandedkar, V. M. (2020). Optimization of railway bogie snubber spring with grasshopper algorithm. In *Vol. 1079. Advances in intelligent systems and computing* (pp. 941–952). Springer. https://doi.org/10.1007/978-981-15-1097-7_80.

Pachpore, S. S., Jadhav, P. V., & Ghorpade, R. R. (2021). Analyzing relation of canal operating force and canal curvature in RCT: A mathematical overview. *Materials Today: Proceedings*, *47*, 5690–5696. https://doi.org/10.1016/j.matpr.2021.03.746.

Patel, S., Kakandikar, G. M., & Kulkarni, O. (2020). Applicability and efficiency of socio-cultural inspired algorithms in optimizing mechanical systems—A critical review. *Review of Computer Engineering Studies*, *7*(2), 31–41. https://doi.org/10.18280/rces.070203.

Yang, X.-S. (2014). In X.-S. Yang (Ed.), *Nature-inspired optimization algorithms*. https://doi.org/10.1016/B978-0-12-416743-8.00011-7.

CHAPTER TEN

A comprehensive review of agriculture irrigation using artificial intelligence for crop production

Maithili Anil Chougule[a] and Anil S. Mashalkar[b]
[a]School of Mechanical Engineering, Dr Vishwanath Karad MIT World Peace University, CAD/CAM/CAE, Pune, India
[b]School of Mechanical, Dr. Vishwanath Karad MIT World Peace University, Pune, India

10.1 Introduction

A lot of agricultural produce gets damaged due to improper guidance related to soil fertility, climatic changes that occur during the process, and the final time of harvesting. The varied topographical extent of our country makes a multitude of harvest timelines and crop varieties. Such complexities and localization make Indian agriculture all the more dynamic and adaptive. Mechanics-backed surveying, mapping, and surveillance tech can help the farmers get better output. Currently, digital farming to automate formerly difficult procedures uses data analysis, sensors, and robotics to autonomous the devices. Mainly focus on the services like self-directed harvesting, field scouting, and weed control. Sensor technology was used in Agriculture 3.0, Information and Communication Technology (ICT) technology was used in Agriculture 4.0, and artificial intelligence and robotics technology was used in Agriculture 5.0.

Object recognition, work planning algorithms, digitalization, and sensor optimization are some of the major enhancements. According to a UN report, the world may face a food supply crisis as a result of the mismatch between population increase and crop output growth. To meet the demands, it is estimated that 2.5 times the current yield will be required (Big Hairy Audacious Goals - BHAG). The term "automation" refers to the method that requires almost no human intervention (Kamilaris & Prenafeta-Boldú, 2018). Machine vision is used in a variety of automated operations in crop processing. Machine vision automation, according to a tech brief, has brought a never-before-seen solution to human problems (Moller, 2010). Image sensors have been used in machine vision to detect and separate materials. Several researchers have examined the use of machine vision to automate agricultural grain and seed process management (Abdullahi et al., 2015). Agriculture is crucial to the long-term viability of any economy. Although the function it plays varies in every country, it is essential for future economic growth and structural restructuring (Mogili & Deepak, 2018; Shah et al., 2019). Crop production was formerly the sole purpose of agricultural activities. It has expanded to covercrop and livestock processing, production, marketing, and distribution in the last two decades. Agricultural activities are now helping to expand the economy as a whole by raising GDP, serving as a source of state trade, lowering unemployment, and supplying raw materials for other sectors (Food and Agriculture Organization of the United Nations, 2017). Artificial intelligence is an emerging technology science that integrates information sciences, psychology, thinking science, systems science, cognitive science, and biotechnology science to research and development of application systems for replicating the system and extending human intelligence. Artificial intelligence is a simulation of the processes of data interaction in human intelligent think, to understand the essence of human intelligence thinking and then design a smart computer that can respond and solve problems in the same way that individuals do (Jani et al., 2020; Parekh et al., 2020).

Due to the use of Agriculture 5.0 technologies like data analysis, integrated robots, and sensors to move from laborious operations to continually mechanized activities, digital farming is becoming increasingly popular. This study examines recent achievements in agricultural-embedded robotics, namely, those employed in autonomous irrigation systems, weeding operations, crop monitoring processes, and chemical spraying processes (Kim et al., 2008). Multi-robots, human–robot cooperation, and environment reconstruction using ground-based sensors for the building of virtual farms were all mentioned as some of the digital farming gateways (Ahir et al., 2020; Gandhi et al., 2020; Kundalia et al., 2020). One of the emerging trends and areas of research in agricultural field robotics is the creation of a swarm of small-scale robots that work together to maximize farming inputs and expose hidden or suppressed data (Murugesan & Sivarajan., 2018; Shamshir, 2018).

Combining artificial intelligence's real-time machine learning technology with the exploration skills of unmanned mechanics might help deliver more information about crop health and its surroundings. Until recently, the robots could only show what their cameras had collected. They can now comprehend their environment thanks to artificial intelligence, which allows them to map places, monitor things, and deliver analytical feedback in real-time (Yang, Liusheng, & Hongli Junmin, 2007)!

10.2 Influence of artificial intelligence technology on an agriculture irrigation system

Artificial intelligence-based technologies aid in the improvement of efficiency in all domains and the management of difficulties encountered by numerous businesses, including crop yield, irrigation, soil content sensing, crop monitoring, weeding, and crop establishment in the agricultural sector (Kim et al., 2008).

Using artificial intelligence, embedded robots are constructed in the agricultural area to carry highly valuable artificial intelligence applications. The agricultural area is facing a dilemma as a result of the expanding global population, but artificial intelligence can provide a better answer. Farmers have been enabled by artificial intelligence-based technology solutions to create higher-quality output while also providing a faster time-to-market crop yield. By 2050, the average farm is expected to produce 4.1 million data points each day.

10.3 Embedded robotics and autonomous in agriculture

Large segments of the economy with low productivity, such as agrifood, are implementing robotics and autonomous systems. According to UK-RAS White Papers, the UK Agri-Food chain generates £108 billion in annual revenue and employs 3.7 million people in a global industry that produced £20 billion in exports in 2016 (2018). Agriculture production and management have benefited greatly from robotics.

Because traditional farming gear is inefficient, researchers have begun to focus on technology to build autonomous agricultural implements (Dursun & Ozden, 2011). The major goal of developing this technology is to replace human labor and provide effective advantages in both small- and large-scale manufacturing (Manivannan & Priyadharshini, 2016). Robotic technology has greatly increased production in this industry (Pedersen et al., 2008).

The embedded robot's execution several agricultural processes freely for irrigation, weeding. Also, protecting the farms from adverse environmental conditions and ensuring the production, effective reports, increasing precision, and managing every plant (Ahir et al., 2020; Gandhi et al., 2020; Kundalia et al., 2020).

In 1794, Eli Whitney (1765–1825) invented the machine, which revolutionized cotton manufacture by considerably accelerating the process of extracting seed from cotton fiber. It generated 50 pounds of cotton in a single day. As a result, self-driving agricultural robots were created.

To determine the real location of seeds, a rudimentary automated model was devised. Seed planting with extreme accuracy was also achieved. Mechanisms for ensuring that the seeds placed have no ground velocity. The seed must not migrate from its initial place following contact with the soil. The plant's whole growth data were automatically stored (Griepentrog et al., 2005).

Various biosensors have been developed to track plant development and detect disease (Tothill, 2001). The human weeding technique was replaced with laser welding technology, in which a mobile-focused infrared light breaks the weeds' cells and is controlled by computers. To make optimum use of water, automated irrigation systems were also implemented (Griepentrog et al., 2006).

10.4 Smart irrigation systems in agriculture

Agriculture consumes 85% of all available freshwater resources on the planet. And this proportion is growing in lockstep with population growth and increased food consumption. As a result, more effective approaches are needed to ensure that water resources are used efficiently in irrigation. Manual watering based on soil water monitoring has been replaced by automatic irrigation scheduling systems. Plant evapotranspiration was taken into consideration while adopting autonomous irrigation equipment, which was dependent on numerous environmental parameters, including humidity, wind speed, solar radiation, and even crop elements like the stage of growth, plant density, soil qualities, and pest.

Machine-to-machine (M2M) technology has been created to make communication and data exchange between nodes in the agricultural field easier, as well as to the server or cloud, via the main network. They also created an automated robotic model for detecting moisture content and temperature in Arduino and Raspberry Pi3 boards. The data is read and updated on a regular basis by the Arduino microcontroller, which also converts analog to digital data. The sign is the Raspberry Pi3, which sends a signal to the Arduino to switch on and off the irrigation water pump. The water will be delivered to the field, and the sensor value will be recorded (Shekhar et al., 2017). For decreasing personnel and time consumption in the irrigation process, an automated irrigation system using Arduino technology was developed (Jha et al., 2019). Developed the concept of a highly efficient and automated watering system based on remote sensors and Arduino technology, which may improve production by up to 40% (Savitha, 2018). Another automated watering system is used in these activities, which includes a soil moisture sensor to detect moisture in the soil, a temperature sensor to read the soil temperature, a pressure regulator sensor to keep the pressure in the soil constant, and a molecular sensor to track crop growth. In addition, digital cameras will be installed. Using a wireless network hotspot, all of these devices' output is transformed into a digital signal, which is then sent to the multiplexer. Because it is immediately placed beneath the crop, the subsurface drip irrigation method avoided water loss due to evaporation and runoff. In a subsequent step, the researchers utilize sensors to read the soil moisture of the fields, as well as raindrop sensors. This system is powered by a wireless network and solar energy. The farmer gets the SMS reading of soil moisture and raindrop sensor by cell phone of GSM module. So, farmers can take the call for ON and OFF the water supply. This will easier for the farmer to take the call for the ON and OFF of the water supply in the field in raining or dry weather conditions (Varatharajalu & Ramprabu, 2018).

Several technics are used to read the moisture content but a soil moisture sensor is one of them. It is buried near the crop's root zones. Sensors provide a precise readout of the moisture content, which is transmitted to the irrigation controller. Soil moisture sensors can also help you save a lot of water (Dukes et al., 2009; Quails et al., 2001).

The moisture sensor is one technique in which farmers can set the threshold of moisture according to the soil capacity, and only when the water required or moisture content is below the threshold value, sensors permit the controller to watering will be allowed in that zone. When the usual watering time approaches, it will first scan the zone's moisture threshold value and then authorise irrigation based on demand. For each zone, the start time and duration are required (Yong et al., 2018).

10.4.1 Dielectric method of moisture content determination

The moisture content of soil can be measured using the soil mass permittivity (dielectric constant) (Gebregiorgis & Savage, 2006). The novel technique presents an automated system for real-time irrigation control that incorporates dielectric soil moisture sensors (Kuyper & Balendonck, 2001). The dielectric property-based measuring approach is thought to be the most promising (Xing et al., 2010). However, tensiometers and electrical resistance impact the accuracy of dielectric capability for transfering electricity (Hanson et al., 2000). Soil has a constant dielectric capability for transferring electricity. Because the soil comprises many components such as air, water, and minerals, the overall commitment of each of these sections is used to compute the dielectric constant.

The capacity of soil to carry power or electricity is the only dielectric stable property. Because soil is made up of several components such as minerals, air, and water, the overall commitment of each of these segments influences the estimation of its dielectric constant. Because the assessment of the dielectric value of water (Kaw = 81) is significantly bigger than the estimation of this consistency for the other soil sections, the estimated value of permittivity is mostly represented by the vicinity of moisture in the soil. One technique to calculate the relationship between the dielectric constant (Kab) and volumetric soil moisture (VWC) is to use Topp et al.'s equation:

$$VWC = -5.3 * 10^{-2} + 2.29 * 10^{-2} Ka_b - 5.5 * 10^{-4} Ka_{b^2} + 4.3 * 10^{-6} Ka_{b^3}$$

Another method for determining the dielectric constant is time domain reflectometry (TDR). The time it takes an electromagnetic wave to travel along a transmission line surrounded by dirt is used to calculate it. The propagation velocity (V) is roughly similar to the square of the transmission time (t in a flash) down and back along the transmission line since it is a component of the dielectric constant (Kab):

$$K_{ab} = (c/v)^2 = ((ct)/(2L))^2$$

10.4.2 Neutron moderation management systems

This is an alternative method for calculating the moisture content of the soil. Fast neutrons are fired from a disintegrating radio dynamic source, such as 241 Am/9Be, and when they collide with particles of equal mass (protons, H+), they slow down dramatically, forming a "cloud" of "thermalized" neutrons. Given that water is the principal source of hydrogen in soil, the thickness of thermalized neutrons surrounding the test often correlates with the division of water in the soil. The test, which is set up as a long and narrow chamber, uses a source and a locator. In this test, estimates are acquired by inserting the test in an entrance tube that has been previously buried in the soil. By balancing the test in the cylinder at various depths, one may determine the amount of moisture in the soil. A direct alignment between the check speed of thermalized neutrons read from the test and the soil moisture content collected from neighboring field tests is used to calculate the moisture content of the soil (Long & French, 1967).

Sensor location is crucial for irrigation robotics to work properly. Irrigation in different field zones might be controlled by a single sensor. It is also feasible to irrigate separate zones using several sensors. In the first scenario, a single sensor is used to irrigate many zones, and the sensor is put in the driest zone, or the zone that requires the most irrigation, to ensure that the entire field is properly irrigated. Sensors should be placed in the root zone of the plants, where water is extracted (ensuring that there are no air gaps around the sensor). This will ensure that the crops are adequately watered. The SMS controller will then need to be connected to the sensor. Once the sensor replies, the controller will take control of the process.

Once you've made this link, you'll need to choose a soil water threshold. The area where the sensor is buried is then sprayed with water and left for a day. As previously indicated, the water content is now the sensor's threshold for scheduled watering.

After the sensors have acquired the data, the microcontrollers take over. It's a crucial part of the automated irrigation system as a whole. A transformer, a bridge rectifier circuit (a portion of an electronic power supply that rectifies AC input to DC output), and a voltage regulator provide 5 V to the whole circuit. After that, the microcontroller is set up. The signals from the sensors are received by the microcontroller. The OP-AMP acts as a link between the sensors and the microcontroller, allowing the measured soil conditions to be communicated. As a result, irrigator pumps rely on data about soil quality at the time of operation to work. As a result, moisture sensors and microcontrollers may be employed to automate the watering process (Rajpal, Jain, Khare, & Shukla, 2011).

10.5 Weeding management systems

Weeds are non-native plant species that limit or degrade the quality of agricultural crop production (Tu et al., 2001). Weeds compete against crops for nutrients, water, sunlight, etc. and significantly reduce the crop yield. The land is infested with inter- and intra-row weeds in row crops. Weeds in the intra-row zone have been shown to reduce crop production by up to 33% (Knezevic, 2002; Peruzzi et al., 2007). As a result, effective, timely, and well-planned management strategies are essential. In the inter-row zone, there are various methods available for successful weed management, but weed control in the intra-row zone remains a considerable challenge.

Manual weed management is the most effective, but it requires a lot of time and money. Furthermore, manual processes demand frequent human bending, as well as exposure to harmful weed species on occasion. Overall, manual procedures are hazardous to one's health and have been abandoned in the most industrialized countries (Chandel et al., 2018; Chethan, Chander, & Kumar, 2018; Chethan, Tewari, et al., 2018; Tewari et al., 1993; Tewari, Ashok Kumar, et al., 2014; Tewari, Nare, et al., 2014; Tu et al., 2001; Van Der Weide et al., 2008).

Thomas K. Pavlyuchenko, a pioneer weed experimentalist, conducted research on plant competition, as reported in "A History of Weed Science in the United States." After doing extensive research, he came to the conclusion that plant water competition began when their roots in the soil overlap to gather water and nutrients, and that weeds were the most aggressive water competitors. The amount of pounds of water required to create one pound of dry matter is the water demand for the aerial sections of the plant (Zimdahl, 2010). To reach maturity, wild mustard (*Brassica kaber* var. *pinnatifida*) takes four times the water of a fully developed oat plant, whereas common ragweed (*Ambrosia artemisiifolia*) requires three times the water of a maize plant. The water need per acre is calculated by multiplying the plant's production in pounds of dry matter per acre by the plant's water requirement. Plants must also have access to light in order to grow. Tall weeds usually restrict the passage of light to the plants. Shade-intolerant weeds such as green foxtail and redroot pigweed do occur, as do shade-tolerant weeds such as field bindweed, common milkweed, spotted spurge, and Arkansas rose. India loses more than $11 billion in agricultural commodities each year owing to weeds, according to research conducted by specialists at the Indian Council for Agricultural Research, which is more than the Centre's financial allocation for agriculture for 2017–18. As a result, removing these weeds from the fields is critical; otherwise, they will not only take up valuable land space but also impede the growth of other plants (Bak & Jakobsen, 2004).

A vision-based weed identification technique in natural illumination was developed using color picture segmentation and a genetic algorithm for in-field weed sensing. It was created by identifying a site in the hue-saturation-intensity (HSI) shading space (GAHSI) for weed identification in open-air fields utilizing heredity calculation. It takes advantage of unique lightning circumstances like radiant and shady, which were mosaicked to test whether GAHSI could be utilized to locate zones in the field in shading space when these two borders are displayed at the same time. The GAHSI provided these as evidence of the location's existence and severability. The GAHSI execution was computed by comparing the GAHSI-portioned picture to a comparison hand-sectioned reference photo. The GAHSI functioned similarly in this scenario (Tang et al., 2000).

We must distinguish between crop seedlings and weeds before designing an automated weed management system (Bhagyalaxmi et al., 2016; Chang & Lin, 2018). A technique was used to distinguish carrot seedlings from ryegrass seedlings. A novel method used a basic morphological characteristic assessment of leaf shape to accomplish this strategy. By assessing the variance in leaf size, this approach offers a range of performance between 52% and 75% for distinguishing between plants and weeds. Digital imagery was used to construct another weeding procedure. A self-organizing neural network was used in this concept. However, for commercial reasons, this technique did not give the desired results, and it was revealed that an NN-based system already existed that could discern species differences with an accuracy of over 75% (Aitkenhead et al., 2003).

Many automated technologies have been created in the modern world, but previous physical methods that relied on physical interaction with the weeds were utilized. Robotic weeding is dependent on the location and number of weeds. Intra-row weeding was done with traditional spring or duck foot tines, which broke the soil and the root contact with tillage, facilitating weed witling. Tillage, on the other hand, is not recommended since it might destroy the crop-soil interface (Griepentrog et al., 2006). As a result, no-contact approaches such as laser treatments and microspraying, which do not influence root-soil contact, were developed (Heisel et al., 2001). Nakai and Yamada (2014) described how to employ agricultural robots to eradicate weeds and create ways for regulating robot postures in the field. The weeds were suppressed and the robot's posture was controlled using the laser range fielder (LRF) approach. Various visual systems were incorporated into the robot. The first was a gray-level vision, which was used to build a row structure to guide the robot through the rows, and the second was a color-based vision, which was the most important and was used to differentiate a single weed from the rest. A unique algorithm was used to create the row recognition system, which has a precision of 2 cm. The initial experiment of this technique was in a greenhouse, where it was used to manage weeds inside a row of crops (Fennimore et al., 2016). Using vision-based technology, the robots were moved along the row structure to clear weeds and discern the single crop from the weed plants.

10.6 Conclusion

Agricultural technology has evolved over time, and technological advancements have influenced the farming industry in a number of ways. Farming is the major industry in various countries throughout the world, and as the world's population grows, from 7.5 billion to 9.7 billion by 2050, there will be increased strain on land, since just 4% of the land will be farmed, according to UN estimates. As a result, farmers will have to do more with less. Food production will have to expand by 60% to feed an additional two billion people, according to the same report. Traditional methods, on the other hand, are unable to meet this enormous need. Farmers and agro-businesses are being pushed to develop innovative methods to improve output while reducing waste. As a result, artificial intelligence is becoming an increasingly important aspect of the agricultural industry's technological progress. To feed an additional two billion people by 2050, the task is to boost global food output by 50%. Artificial intelligence-driven solutions will not only help agriculturalists increase productivity, but they will also boost crop yield, quality, and speed up a time to market.

The agricultural business has several obstacles, including a lack of appropriate irrigation schemes, weeds, and harsh climate situations. However, with the benefit of technology, performance may be better quality, and therefore, these issues can be resolved. It may be enhanced using artificial intelligence-driven approaches such as remote sensors for noticing soil moisture content and GPS-assisted automatic watering. The problem for farmers was that precision weeding techniques were able to offset the massive amount of crops lost during the weeding process. These self-driving robots are not just becoming more prevalent. For starters, man-made brainpower in agriculture challenges may be used to understand resource and employment shortages. In traditional methods, a large amount of labor was necessary to get agricultural parameters such as plant height, soil quality, and content, which necessitated manual testing, which was time-consuming. With the aid of the various systems explored, quick and non-damaging high-throughput phenotyping would be achievable, with the added benefit of flexible and favorable activities, on-demand access to information, and spatial goals.

10.7 Future scope

The young farmer makes more interest and invests in artificial intelligence than the elder farmers. The new technology has been introduced step-by-step with time. Step-by-step agriculture domain is moving toward precise farming, which will manage in individual plant and crop also. For sustainable growth of plant and crop, deep learning and covering other automation methods should be done to provide a favorable atmosphere to the crop. Ultimately, the production of extra crops yields an increase in the variety of products and production techniques. Artificial intelligence methods that use computational network can help to detect disease of crop and unwanted weeding in the crop. Like a greenhouse, we cannot provide the control environment to the crop, but using wireless technology and sensors, we can monitor the weather without human intervention. The farmer can use the robots for spaying, crop weeding, seeding, planting, harvesting, fertilizing, irrigation, and shepherding. The drones with thermal cameras give a continuous real-time data of the farm to start irrigation and prevent the water flooding or crop get the advent amount of water always.

References

Abdullahi, H. S., Mahieddine, F., & Sheriff, R. E. (2015). Technology impact on agricultural productivity: A review of precision agriculture using unmanned aerial vehicles. In *Vol. 154. Lecture notes of the institute for computer sciences, social-informatics and telecommunications engineering, LNICST* (pp. 388–400). Springer Verlag. https://doi.org/10.1007/978-3-319-25479-1_29.

Ahir, K., Govani, K., Gajera, R., & Shah, M. (2020). Application on virtual reality for enhanced education learning, military training and sports. *Augmented Human Research, 5*(1). https://doi.org/10.1007/s41133-019-0025-2.

Aitkenhead, M. J., McDonald, A. J. S., Dawson, J. J., Couper, G., Smart, R. P., Billett, M., et al. (2003). A novel method for training neural networks for time-series prediction in environmental systems. *Ecological Modelling, 162*(1–2), 87–95. https://doi.org/10.1016/S0304-3800(02)00401-5.

Bak, T., & Jakobsen, H. (2004). Agricultural robotic platform with four wheel steering for weed detection. *Biosystems Engineering, 87*(2), 125–136. https://doi.org/10.1016/j.biosystemseng.2003.10.009.

Bhagyalaxmi, K., Jagtap, K. K., Nikam, N. S., Nikam, K. K., & Sutar, S. S. (2016). Agricultural robot (irrigation system, weeding, monitoring of field, disease detection). *International Journal of Innovative Research in Computer and Communication Engineering, 4*(3), 4403–4409.

Chandel, A. K., Tewari, V. K., Kumar, S. P., Nare, B., & Agarwal, A. (2018). On-the-go position sensing and controller predicated contact-type weed eradicator. *Current Science, 114*(7), 1484–1485. https://doi.org/10.18520/cs/v114/i07/1485-1494.

Chang, C. L., & Lin, K. M. (2018). Smart agricultural machine with a computer vision-based weeding and variable-rate irrigation scheme. *Robotics*, 7(3). https://doi.org/10.3390/robotics7030038.

Chethan, C. R., Chander, S., & Kumar, S. P. (2018). Dynamic strength based dryland weeders-ergonomic and performance evaluation. *Indian Journal of Weed Science*, 50(4), 382. https://doi.org/10.5958/0974-8164.2018.00081.3.

Chethan, C. R., Tewari, V. K., Nare, B., & Kumar, S. P. (2018). Transducers for measurement of draft and torque of tractor-implement system—A review. *AMA, Agricultural Mechanization in Asia, Africa and Latin America*, 49(4), 81–87. http://www.shin-norin.co.jp.

Dukes, M. D., Shedd, M., & Cardenas-Lailhacar, B. (2009). *Smart irrigation controllers: How do soil moisture sensor (SMS) irrigation controllers work?* (pp. 1–5). IFAS Extension.

Dursun, M., & Ozden, S. (2011). A wireless application of drip irrigation automation supported by soil moisture sensors. *Scientific Research and Essays*, 6(7), 1573–1582. http://www.academicjournals.org/sre/PDF/pdf2011/4Apr/Dursun%20and%20Ozden.pdf.

Fennimore, S. A., Slaughter, D. C., Siemens, M. C., Leon, R. G., & Saber, M. N. (2016). Technology for automation of weed control in specialty crops. *Weed Technology*, 30(4), 823–837. https://doi.org/10.1614/WT-D-16-00070.1.

Food and Agriculture Organization of the United Nations. (2017). *The State of Food and Agriculture Leveraging Food Systems for Inclusive Rural Transformation* (pp. 1–181). FAO. ISBN: 978-92-5-109873-8.

Gandhi, M., Kamdar, J., & Shah, M. (2020). Preprocessing of non-symmetrical images for edge detection. *Augmented Human Research*, 5(1). https://doi.org/10.1007/s41133-019-0030-5.

Gebregiorgis, M. F., & Savage, M. J. (2006). Determination of the timing and amount of irrigation of winter cover crops with the use of dielectric constant and capacitance soil water content profile methods. *South African Journal of Plant and Soil*, 23(3), 145–151. https://doi.org/10.1080/02571862.2006.10634746.

Griepentrog, H. W., Nørremark, M., Nielsen, H., & Blackmore, B. S. (2005). Seed mapping of sugar beet. *Precision Agriculture*, 6(2), 157–165. https://doi.org/10.1007/s11119-005-1032-5.

Griepentrog, H. W., Nørremark, M., Nielsen, J., & Soriano, J. F. (2006). Close-to-crop thermal weed control using a CO2 laser. In *Proc. proceedings of CIGR world congress* (pp. 3–7).

Hanson, B., Peters, D., & Orloff, S. (2000). Effectiveness of tensiometers and electrical resistance sensors varies with soil conditions. *California Agriculture*, 54(3), 47–50. https://doi.org/10.3733/ca.v054n03p47.

Heisel, T., Schou, J., Christensen, S., & Andreasen, C. (2001). Cutting weeds with a CO 2 laser. *Weed Research*, 41(1), 19–29. https://doi.org/10.1046/j.1365-3180.2001.00212.x.

Jani, K., Chaudhuri, M., Patel, H., & Shah, M. (2020). Machine learning in films: An approach towards automation in film censoring. *Journal of Data, Information and Management*, 2(1), 55–64. https://doi.org/10.1007/s42488-019-00016-9.

Jha, K., Doshi, A., Patel, P., & Shah, M. (2019). A comprehensive review on automation in agriculture using artificial intelligence. *Artificial Intelligence in Agriculture*, 2, 1–12. https://doi.org/10.1016/j.aiia.2019.05.004.

Kamilaris, A., & Prenafeta-Boldú, F. X. (2018). Deep learning in agriculture: A survey. *Computers and Electronics in Agriculture*, 147, 70–90. https://doi.org/10.1016/j.compag.2018.02.016.

Kim, Y., Evans, R. G., & Iversen, W. M. (2008). Remote sensing and control of an irrigation system using a distributed wireless sensor network. *IEEE Transactions on Instrumentation and Measurement*, 57(7), 1379–1387. https://doi.org/10.1109/TIM.2008.917198.

Knezevic, S. Z. (2002). *The concept of critical period of weed control cooperative extension* (pp. 30–40).

Kundalia, K., Patel, Y., & Shah, M. (2020). Multi-label movie genre detection from a movie poster using knowledge transfer learning. *Augmented Human Research, 5*(1). https://doi.org/10.1007/s41133-019-0029-y.

Kuyper, M. C., & Balendonck, J. (2001). Application of dielectric soil moisture sensors for real-time automated irrigation control. In *Vol. 562. Acta horticulturae* (pp. 71–79). International Society for Horticultural Science. https://doi.org/10.17660/ActaHortic.2001.562.7.

Long, I. F., & French, B. K. (1967). Measurement of soil moisture in the field by neutron moderation. *Journal of Soil Science, 18*(1), 149–166. https://doi.org/10.1111/j.1365-2389.1967.tb01496.x.

Manivannan, L., & Priyadharshini, M. S. (2016). Agricultural robot. *International Journal of Advanced Research in Electrical, Electronics and Instrumentation Engineering,* 153–156.

Mogili, U. R., & Deepak, B. B. V. L. (2018). Review on application of drone systems in precision agriculture. In *Vol. 133. Procedia computer science* (pp. 502–509). Elsevier B.V. https://doi.org/10.1016/j.procs.2018.07.063.

Moller, J. (2010). Computer vision—A versatile technology in automation of agriculture machinery. In *21st Annual meeting* (pp. 1–16).

Murugesan, R., & Sivarajan., S. (2018). Industry 4.0 for sustainable development. *Annual technical volume of Mechanical Engineering Division Board, Vol. 3.*

Nakai, S., & Yamada, Y. (2014). Development of a weed suppression robot for rice cultivation: Weed suppression and posture control. *Computer, Electronics and Communication Engineering, 8*(12), 1658–1662.

Parekh, V., Shah, D., & Shah, M. (2020). Fatigue detection using artificial intelligence framework. *Augmented Human Research, 5*(1). https://doi.org/10.1007/s41133-019-0023-4.

Pedersen, S. M., Fountas, S., & Blackmore, S. (2008). Agricultural robots—Applications and economic perspectives. In *Service Robot Applications* (pp. 369–382).

Peruzzi, A., Ginanni, M., Raffaelli, M., & Fontanelli, M. (2007). Physical weed control in organic fennel cultivated in the Fucino Valley. In *Proceedings of the 7th workshop of the EWRS working group on physical and cultural weed control* (pp. 32–40).

Quails, R. J., Scott, J. M., & DeOreo, W. B. (2001). Soil moisture sensors for urban landscape irrigation: Effectiveness and reliability. *Journal of the American Water Resources Association, 37*(3), 547–559. https://doi.org/10.1111/j.1752-1688.2001.tb05492.x.

Rajpal, A., Jain, S., Khare, N., & Shukla, A. K. (2011). Microcontroller-based automatic irrigation system with moisture sensors. *Proceedings of the International Conference on Science and Engineering,* 94–96.

Savitha, M. (2018). Smart crop field irrigation in IoT architecture using sensors. *International Journal of Advanced Research in Computer Science, 9*(1), 302–306. https://doi.org/10.26483/ijarcs.v9i1.5348.

Shah, G., Shah, A., & Shah, M. (2019). Panacea of challenges in real-world application of big data analytics in healthcare sector. *Journal of Data, Information and Management, 1*(3–4), 107–116. https://doi.org/10.1007/s42488-019-00010-1.

Shamshir, R. R., et al. (2018). Research and development in agricultural robotics: A perspective of digital farming. *International Journal of Agricultural and Biological Engineering, 11*(4).

Shekhar, Y., Dagur, E., Mishra, S., Tom, R. J., Veeramanikandan, M., & Sankaranarayanan, S. (2017). Intelligent IoT based automated irrigation system. *International Journal of Applied Engineering Research, 12*(18), 7306–7320. http://www.ripublication.com/ijaer.htm.

Tang, L., Tian, L., & Steward, B. L. (2000). Color image segmentation with genetic algorithm for in-field weed sensing. *Transactions of the American Society of Agricultural Engineers, 43*(4), 1019–1027.

Tewari, V. K., Ashok Kumar, A., Nare, B., Prakash, S., & Tyagi, A. (2014). Microcontroller based roller contact type herbicide applicator for weed control under row crops. *Computers and Electronics in Agriculture*, *104*, 40–45. https://doi.org/10.1016/j. compag.2014.03.005.

Tewari, V. K., Datta, R. K., & Murthy, A. S. R. (1993). Field performance of weeding blades of a manually operated push-pull weeder. *Journal of Agricultural Engineering Research*, *55* (2), 129–141. https://doi.org/10.1006/jaer.1993.1038.

Tewari, K., Nare, B., Kumar, S. P., Chandel, A. K., & Tyagi, A. (2014). A six-row tractor mounted microprocessor-based herbicide applicator for weed control in row crops. *International Pest Control*, *56*(3), 162–164.

Tothill, I. E. (2001). Biosensors developments and potential applications in the agricultural diagnosis sector. *Computers and Electronics in Agriculture*, *30*(1–3), 205–218. https://doi. org/10.1016/S0168-1699(00)00165-4.

Tu, M., Callie, H., & John, M. R. (2001). Tools and techniques for use in natural areas. *Weed Control Methods Handbook*. The Nature Conservancy.

Van Der Weide, R. Y., Bleeker, P. O., Achten, V. T. J. M., Lotz, L. A. P., Fogelberg, F., & Melander, B. (2008). Innovation in mechanical weed control in crop rows. *Weed Research*, *48*(3), 215–224. https://doi.org/10.1111/j.1365-3180.2008.00629.x.

Varatharajalu, K., & Ramprabu, J. (2018). Wireless irrigation system via phone call & sms. *International Journal of Engineering and Advanced Technology*, *8*(2), 397–401. https://www. ijeat.org/wp-content/uploads/papers/v8i2s/B10821282S18.pdf.

Xing, Z., Zheng, W., Shen, C., Yang, Q., & Sun, G. (2010). The measurement of soil water content using the dielectric method. In *2010 World automation congress, WAC 2010* (pp. 241–245).

Yang, H., Liusheng, W., & Hongli Junmin, X. (2007). Wireless sensor networks for intensive irrigated agriculture. *Proceedings of the 4th IEEE Consumer Communications and Networking Conference*, 197–201.

Yong, W., Shuaishuai, L., Li, L., Minzan, L., Ming, L., Arvanitis, K. G., et al. (2018). Smart sensors from ground to cloud and web intelligence. *IFAC-PapersOnLine*, *51*(17), 31–38. https://doi.org/10.1016/j.ifacol.2018.08.057.

Zimdahl, R. L. (2010). Ethics for weed science. *Pakistan Journal of Weed Science Research*, *16* (2), 109–121.

Index

Note: Page numbers followed by *f* indicate figures and *t* indicate tables.

Printed in the United States
by Baker & Taylor Publisher Services